Climate Change Explored and Explained

Published by Stephen Harvie

Edition 11 - June 2013.

ISBN 978-0-646-53513-5.

About the author

Stephen Harvie is a citizen of Australia. He became interested in the climate change debate in February 2010 after reading the Australian skeptic Ian Plimer's book, "*Heaven + Earth – Global Warming: The Missing Science*" whilst staying with friends Lesley and Terry on Australia's south east coast. Reading that book prompted the writing of this one.

This book is the result of sifting through tens of thousands of pages of scientific, skeptical and government policy material in order to decipher both the scientific and political climate change debate. This book summarises the results of that research, in two parts – one for the science and one for the policies.

This book discusses policies for the world, but focuses particularly on policies proposed in Australia, including emissions trading, carbon tax and direct action policies. It contains an analysis of current policies of Australian political parties and is therefore a political statement. It also discusses proposed policies in the USA. The author is not a member of any political party and does not have any political affiliations.

About Australia

Australia is one of the world's highest per capita carbon emitting nations, just ahead of the USA.

The current Australian Government originally proposed that Australia implement an emissions trading scheme as an approach to reducing carbon emissions. As an alternative approach, Australia's current Opposition party has proposed a low-cost direct-action approach. Both of these policy proposals currently only target modest emission reductions of 5% by 2020.

Australia is the world's largest exporter of coal. Coal is Australia's largest export. In coming years Australia's economy is expected to grow at around 3% p.a. resulting from a resources and mining boom, based largely on fossil fuels such as coal, oil and gas.

Cover picture. Terry wading on the road somewhere near the bridge over a local creek, from the photo album "Living and Loving Climate Change" by Lesley. Prior to this deluge, this area of Australia had been in a state of drought. They live in the red patch on the bottom right of the map below.

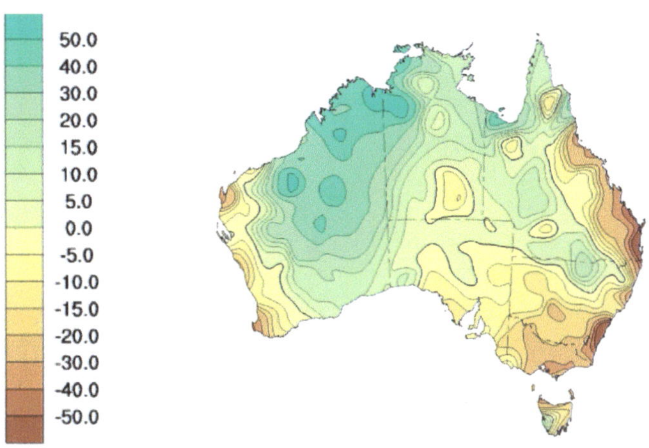

Trend in Australian annual rainfall 1960-2009 mm per decade. Source: Bureau of Meteorology.

First letter

21 July 2010

The Honourable Julia Gillard MP
Prime Minister of Australia
PO Box 6022
House of Representatives
Parliament House
Canberra ACT 2600

Dear Ms Gillard

Please accept as a gift the enclosed book that I have recently written about the very important topic of climate change.

Significant ideas that the book contains include:

- A discussion of why carbon-pricing based policies, used alone, might not be efficient or effective and might not work.
- An alternative large-scale direct-action policy, with more certainty of solving the problem of climate change, with costed plans for Australia and the USA to be carbon neutral by 2040.
- Costed plans for Australia, the USA, China and the world implement 100% renewable energy by 2040.
- Recommendations for the major Australian political parties for how to make their climate change policies more palatable to voters.
- Recommendations for the basis of a new world climate change accord.

As well as putting a price on carbon, the book advocates for Australia a very large-scale direct-action plan costing $13B per annum for 30 years. This is around 1% of GDP. This is important to protect our future. I'd be happy to discuss the book further.

Yours sincerely

Stephen Harvie

Seventh letter

3 December 2012

Senator the Hon Bob Carr

Level 10, Bligh House,

4-6 Bligh Street

Sydney, NSW, 2000

Dear Mr Carr

I would like to provide you with this electronic edition of my book Climate Change Explored and Explained, on the enclosed CD.

It contains a climate policy proposal for Australia which much more effective and efficient than our current emissions trading scheme. This includes plans and costings for Australia to achieve 100% renewable energy and be carbon neutral in 30 years.

It also goes further and includes a high level design and costing for a world 100% renewable energy system, in which Australia plays a major part in generating 10% of the world's renewable electrical energy needs. The predicted profits of building this are many times Australia's current GDP.

I do believe the proposed world renewable energy system needs to be built, as each country cannot solve the problem of climate change alone.

Participation in this world system should be open to any country that is either capable of generating more than its energy requirements, or is willing to participate in joint venture arrangements with other countries that can, like Australia.

Australia and the USA could work together to begin this process. I would also encourage discussion of this topic with China.

I would encourage you to look at these ideas and consider what might be done to help bring them about.

Yours sincerely

Stephen Harvie

Second letter

19 April 2011

The Hon Julia Gillard MP
Prime Minister of Australia
PO Box 6022
Parliament House
Canberra ACT 2600

Financial Savings of Rapid Renewable Energy Implementation and Potential for a Low Carbon Price

Dear Ms Gillard

Please find enclosed the most recent edition of a book I have written about climate change policy and science, titled "*Climate Change Explored and Explained*". I also wish to provide you with this very important update to the 30 year climate change mitigation plan that is presented in this book.

Attached is a summary of costs of rapidly implementing renewable energy vs. retaining coal.

This new analysis shows that implementing 100% renewable energy over 30 years is profitable compared to staying with coal. The 30 year savings are summarised below, based on different starting coal prices, and with and without future discounting.

Coal Price	Renewable Energy Implementation Cost	Savings from Reduced Coal	Plan Profit Over 30 Years
Starting $50/tonne (historical price pre 2007)	$229B ($126B [1])	$506B ($203B [1])	$277B ($78B [1])
Starting $120/tonne (typical post 2007 price)	$229B ($126B [1])	$1,025B ($386B [1])	$797B ($260B [1])

1. The figures in parentheses are future discounted at 5% p.a.

Climate Change Explored and Explained describes a 30 year plan for Australia to become carbon neutral across all emission sectors, costing $390B. It notes there would be reduced running costs by converting from coal to renewable energy, but does not analyse them. This update shows the proposed plan would not have a $390B cost compared to business-as-usual. It would be profitable and self-funding.

I can provide this update to my book after carefully examining and checking the excellent work of *Beyond Zero Emissions* in their 10 year renewable energy implementation plan. I also referred to current and forecast renewable energy cost data from the European Commission's 2008 NEEDS analysis, Sargent and Lundy and Lazard.

The *Beyond Zero Emissions* plan shows that rapidly implementing 100% renewable energy in 10 years is similar in cost over a 30 year period to retaining coal. That is, savings from reduced coal usage over 30 years matched their proposed 10 year implementation cost. They also showed further savings of $1 trillion dollars if public transportation is also converted to electrical energy.

My 30 year plan is more conservative than that of Beyond Zero Emissions. It has a longer implementation timeframe. My analysis assumes less progress on future energy efficiency than theirs. It also assumes slower reductions in renewable energy prices, but takes more advantage of them over a longer implementation timeframe. The analysis shows greater expected profits. I do not also analyse the further savings from converting public transportation to renewable energy.

Core Conclusions

The *Beyond Zero Emissions* 10 year plan and my new 30 year analysis thus show that a strong plan for conversion to renewable energy will be profitable. This is without and before any additional price on carbon. This is due to rising fossil fuel prices and falling renewable energy prices. The additional potential impact of a carbon price on the cost of coal is also shown separately in my attached analysis.

The attached analysis shows that the 30 year cost of renewable energy is cheaper than coal because the 30 year capital cost of renewable energy is not much more than coal, but running costs are far less. The analysis shows a 30 year capital price of renewable energy of $2.2M per megawatt. This is only slightly higher than the capital cost of coal power, for which I have assumed no increase in cost over the time period. The significant savings come from reduced fossil fuel usage, with savings growing to $100B per annum at the end of the 30 year implementation.

In order of size, the analysis shows that the greatest effects on the overall price differential are firstly increasing fossil fuel prices, secondly a carbon price (if applied), and lastly falling renewable energy prices. I think this is the reverse to what is generally thought. The result will vary depending on what happens with each of them.

The expected size of fossil fuel and renewable energy price changes means that a carbon price could be kept small, as a means of providing an ongoing incentive for energy efficiency across other emission sectors besides electricity generation. Some more analysis of this is provided on the following page.

Because of their profitability, the plans advocated by Beyond Zero Emissions and/or by myself make economic sense, regardless of whether one believes in the science of climate change, and therefore have potential to unify public opinion. As the Beyond Zero Emissions plan shows, more renewable energy jobs would be created than lost from coal. If we increased our coal exports in line with the reduced onshore usage, no jobs would be lost.

The recent 2011 Garnaut review update emphasised that additional investment in the first 10 years of around $2.5B p.a. as well as having a carbon price will be necessary, in order to deal with the expected "market failures". This amount suggested for the first 10 years is greater than the annual net cost of the plan in its first 10 years after deduction of coal savings. The plan is profitable on an annual basis after that.

One common fear of carbon pricing is that significant renewable energy and new green jobs will not be created until the carbon price becomes sufficiently high. This lack of confidence results from not only uncertainty of the future carbon price but also of the speed of reduction of renewable energy prices and expected increases in fossil fuel prices. The carbon price is under much discussion but these other potential price movements are not. These other factors will have the bigger effect.

As a result of these uncertainties there is expected to be a significant time delay between the carbon price starting and new jobs being created. Some industries expect to experience job losses. Without the adoption of a well defined and strong long term renewable energy implementation plan, I believe this will be true. Again, this is probably because of the uncertainties, a lack of knowledge and confidence, as well as the lack of a strong, specific and far-sighted plan for a nationally coordinated approach to large-scale renewable energy implementation.

The Cost of Delay

Intuitively, one might first think that delaying implementation until renewable energy prices become cheaper would be the least cost approach, but my analysis shows this is not the case. The table below summarises costs of the 30 year implementation plan, assuming a prompt start, and showing its increased benefits from coal savings extended out to 40 years.

Coal Price	Renewable Energy Implementation Cost	Savings from Reduced Coal	Plan Profit Over 40 Years
Starting $50/tonne (historical price pre 2007)	$242B ($127B [1])	$1,150B ($321B [1])	$909B ($194B [1])
Starting $120/tonne (typical post 2007 price)	$242B ($127B [1])	$2,509B ($657B [1])	$2,267B ($530B [1])

In contrast, the table below summarises the costs and benefits of the 30 year implementation plan if decision to implement is delayed by 10 years, matching the same 40 year timeframe as above.

Coal Price	Renewable Energy Implementation Cost	Savings from Reduced Coal	Plan Profit Over 40 years
Starting $50/tonne (historical price pre 2007)	$177B ($57B [1])	$779B ($188B [1])	$622B ($131B [1])
Starting $120/tonne (typical post 2007 price)	$177B ($57B [1])	$1,730B ($389B [1])	$1,553B ($332B [1])

1. The figures in parentheses are future discounted at 5% p.a.

It can be seen from this comparison that there are economic benefits as well as environmental benefits of starting earlier. Although renewable energy capital costs start high, these are dwarfed by the probable increased fossil fuel costs as we move down the track. This analysis does not include other significant benefits of starting early, such as faster job creation, and greater potential for export of expertise, technology and green energy.

Sensitivity Analysis

Manipulation of the cost analysis suggests the following results, which are enormously profound.

- The plan remains profitable if the coal price starts at $120/tonne and did not grow for 30 years.
- The plan is break-even with BAU if the coal price starts at $50/tonne and did not grow for 30 years.
- The plan remains profitable if the coal price starts at $120/tonne and grows, renewable costs do not reduce at all for 30 years, and still with no carbon price included.
- The plan is breakeven with BAU if the coal price starts at $120/tonne and does not grow, renewable energy costs do not reduce at all for 30 years, and still with no carbon price.
- The plan is breakeven with BAU if the coal price starts at $50/tonne and does not grow, renewable energy costs do not reduce at all for 30 years, and a carbon price starting at $15/tonne is included.
- The plan is breakeven with BAU with a coal price starting at only $10 per tonne, but growing at 6% p.a. over 30 years to $100 per tonne.
- The plan is breakeven with BAU if coal capital and running costs are assumed to be zero and a carbon price starting at $25/tonne is included.

Final Conclusions

I believe that some of the conclusions of this analysis, in terms of cost outcomes, costs of delay and sensitivity analysis are extremely important and ground-breaking. They add to work already done. Implementation of the *Beyond Zero Emissions* 10 year plan, my more conservative 30 year plan or an alternative plan will require significant leadership, and also picking and implementing a cost effective solution such as the one proposed. There will be significant economies of scale, including implementation, support, maintenance and upgrade costs of implementing a relatively uniform renewable energy solution.

In comparison, I suggest that without a detailed and specific implementation plan, emission trading will by itself result in a much slower, less coordinated and less uniform approach that will be much more costly to implement, maintain, support and upgrade. That is a downside of not "picking winners". It is also more difficult to forecast potential profits, particularly if carbon offsets can be purchased overseas. I would invite a detailed comparison of this plan with the expected costs and outcomes of emissions trading.

I have found it difficult to find out what the Australian coal power industry actually pays for coal. The alternative prices used in my analysis are a conservative approximation of recent and historical commercial prices for Australian coal. I understand that coal usage in Australia is subsidised. If so, my analysis might still reflect actual prices being paid, or alternatively might reflect the true cost including subsidies, or the opportunity cost from lost export potential. My sensitivity analysis looked at coal prices starting as low as $10/tonne. This deserves further analysis and consideration.

I would encourage thinking that a 30 year implementation plan could be 3-5 years in the planning, and 25-27 years in the implementation. That could include an immediate start on pilot implementations. By starting on such a plan, the door remains open to accelerating the implementation to better match the *Beyond Zero Emissions* 10 year plan, or as my book suggests, working with industry to continuously evaluate the best financial course and changing direction as required.

The profitability of the plan takes away the scientific debate. Further, I believe that the best international thinking suggests that a combination both direct action and a carbon price provides the best solution. That then takes away the political debate, as would the potential for a lower carbon price. As this analysis suggests, strong action is cost effective even before environmental and risk management benefits are considered.

A lower carbon price will minimise impact on industries other than electricity generation. The plan for electricity generation outlined in this letter, plus the broader plan covering all emissions sections outlined in my book, allows Australia to take substantial action at no net cost.

Please do contact me if you would like to discuss this important topic further, or if you would like me to send you the attached calculation spreadsheets for more expert analysis, verification and improvement.

Yours sincerely

Stephen Harvie

Understanding Concentrating Solar Thermal Energy

- The modelled design is based on an existing more cost-effective (Lazard) technology that uses a large field of mirrors to heat an energy storage fluid at a central tower, typically a "molten salt". The heated fluid is used to generate steam to drive a turbine and generate electricity. The heated fluid is stored to release energy at peak times or the next day.
- A facility is rated by the amount of energy it can generate e.g. 20, 100, 200MW and its capacity factor, the proportion of the year it will generate that energy e.g. 30%, 50% or 70%. Adding energy storage approximately doubles the capital cost of a facility but increases the capacity factor and increases efficiency (NEEDS).
- Both the capital costs and levelised energy costs are projected to reduce over coming years and decades as manufacturing processes and technology improves. The Levelised Energy Cost is expected to reduce over the next decade from over 10 cents per kilowatt hour to around 5 cents per kilowatt hour (NEEDS, Sargent and Lundy).
- The low cost of materials and centralised energy generation of this design allows for gradual construction, future technology upgrade and high local content for construction.
- Modelling of annual weather patterns shows that a geographic spread and backup facilities such as wind and biomass will provide reliable baseload and peak energy generation (Beyond Zero Emissions).
- In other countries plants of 100MW are commencing construction and units of 220 MW are now proposed. A 120 MW plant under construction in California for around US$750M will generate 450 gigawatt hours p.a.. Instead of building a larger 220 MW plant that can generate 1388 gigawatt hours p.a. an alternative would be to build three of the smaller 120 MW plants at a total cost of around $2.2B. This has a lower capital cost per MW of $7.5M/MW, but a similar overall cost to generate a similar amount of energy per year. This helps prove the validity of the proposed starting price of the larger 220MW plant of $10.5M/MW.

Understanding Wind Energy

- A wind tower is rated by the energy it can generate e.g. 2MW, 4MW or more. The capacity of a wind tower is typically about 30%, depending on the amount of wind at its geographic location.
- Capital costs of wind towers are presently in the $1.9-$2.5M/MW range (Lazard). The Levelised Energy Cost of wind is 4.4 to 9.1 cents per kilowatt hour (Lazard).

Comparison with Coal and Gas Energy

- Estimates of the Levelised Energy Cost of coal and gas power generation vary quite widely in the range of 5 cents to over 10 cents per kilowatt hour, depending on the type of power station and also the projected cost of fossil fuels, which are rising.
  Lazard reports current levelised energy costs of pulverised supercritical coal as similar to solar thermal. Some forms of renewable energy are cost competitive with fossil fuels now, and others will become more cost competitive over the coming decade.

References

Levelised Cost of Energy Analysis – Version 2.0, Lazard, June 2008.

Final report on technical data, costs and life cycle inventories of solar thermal power plants, New Energy Externalities Developments for Sustainability (NEEDS), European Commission, 2008.

Assessment of Parabolic Trough and Power Tower Solar Technology Cost and Performance Forecasts, Sargent and Lundy LLC Consulting Group, 2003

Zero Carbon Australia 2020 Stationary Energy Plan, Beyond Zero Emissions, 2010

30 Year 100% Renewable Energy Implementation Plan - $50/tonne Coal Starting Price

General Findings:	
1. Cost of 100% renewable energy rollout	$229 B
2. Savings from reduced coal usage	$506 B
3. Profit of plan over 30 years	$277 B

			CST Part	Wind part
Assumptions	Australia's current electricity usage	257 TWh p.a.		
	Plus 30% population growth	334 TWh p.a.		
	Minus 10% efficiency gain	301 TWh p.a.		
	Starting cost of solar thermal	10.5 $M/MW		
	Starting cost of wind power	2.2 $M/MW		
			CST Part	Wind part
Key Results	Power capacity rating	77 GW	27	50
	Power capacity provided	302 TWh p.a.	171	131
	Cost of implementation	229 $B	170	59
	Average capital cost over 30 years	3.0 $M/MW		
			CST Part	Wind part
Comparison with BZE 10 Year plan	Power capacity rating	93 GW	43	50
	Power capacity provided	325 TWh p.a.	195	131
	Cost of implementation	261 $B	190	71
	Average capital cost over 10 years	2.8 $M/MW		

General notes

1. The BZE 10 year plan assumes greater future efficiency gains, but also costs electrical energy to service conversion of gas heating and transportation to electricity.

2. The BZE 10 year plan also provides $90B for transmission line upgrades and $17B for additional off-grid and backup bio-mass systems.

My book's full 30 year plan funds $255B which therefore covers these latter items. Transmission line upgrades also occur in the BAU case, estimated by BZE at $2B p.a.

3. Total costs therefore depend on overall plan assumptions. This plan is based on current technology. Technology advances over the 30 years will result in a better outcome.

Concentrating solar thermal assumptions

1. As per BZE plan, each 220 MW unit typically creates 1388 GWh p.a. Starting price is $10.5M/MWe, reducing roughly in line with the European Commission's 2008 analysis.
2. This analysis assumes slower CST price reduction over time than the BZE plan, but the longer 30 year timeframe takes greater advantage of reduced prices.

Wind tower assumptions

1. As per BZE plan, assumes 30% efficiency per tower, and similar expected price reductions over first 10 years.

Coal savings assumption

1. 1 Kg coal generates 2kWh electricity, 1MT coal generates 2TWh electricity
2. As estimated by BZE new coal plant capital investment under BAU scenario is $135B over 30 years.
3. This analysis includes capital costs of renewable energy, fuel costs of coal, and capital costs of coal, being the three largest factors.
4. Black coal is the dominant current energy source. Costs vary over time. This analysis assumes start at either $50 or $120/ton and increasing 6% p.a. Coal prices follow oil. These coal prices should be seen as a combined price for a black coal, brown coal, gas fossil fuel mix. Brown coal is cheaper and gas is more expensive, at least historically. Coal prices per tonne and per ton are similar. In reality, fuel cost savings would be offset one year from the analysis shown below.
5. Other operating and maintenance costs of coal or renewable energy are similar and covered in more detail in the BZE analysis and the attached LEC analysis.

	Concentrating Solar Thermal Capital Costs (Annual)						Wind Tower Capital Costs (Annual)						Coal Usage Saved (Annual)				Saved New Coal Plants Capital $B	Total Costs/Savings (Annual)		Plan Cost (savings are negative)		Future Discounting at 5% p.a.			
Year	MWe/unit	Units p.a.	MWe/p.a.	New TWh p.a capacity	$M/MWe	$B	MW/unit	Units p.a.	MW/p.a.	New TWh pa capacity	$M/MW	$B	Coal Twh p.a. Replaced	Coal MT Replaced	Coal Cost/Ton $	Coal Cost saved p.a. $B		Total Costs Renewable Energy $B	Total Coal Savings $B	Annual Net Plan Cost $B	Cumulative Plan Cost $B	Annual Plan Cost $B	Total Coal Savings $B	Annual Net Plan Cost $B	Cumulative Plan Cost $B
1	220	0	0	0.00	10.5	2.31	4	417	1667	4.38	2.2	3.67	4	2	50	0.1	4.5	6.0	4.6	1.4	1.4	6.0	4.6	1.4	1.4
2	220	2	440	2.78	10.5	2.31	4	417	1667	4.38	1.9	3.17	12	6	53	0.3	4.5	5.5	4.8	0.7	2.0	5.2	4.6	0.6	2.0
3	220	3	660	4.16	10.5	6.93	4	417	1667	4.38	1.9	3.17	20	10	56	0.6	4.5	10.1	5.1	5.0	7.1	9.2	4.6	4.6	6.6
4	220	3	660	4.16	9	5.94	4	417	1667	4.38	1.65	2.75	29	14	60	0.9	4.5	8.7	5.4	3.3	10.4	7.5	4.6	2.9	9.5
5	220	3	660	4.16	9	5.94	4	417	1667	4.38	1.65	2.75	37	19	63	1.2	4.5	8.7	5.7	3.0	13.4	7.1	4.7	2.5	11.9
6	220	3	660	4.16	9	5.94	4	417	1667	4.38	1.25	2.08	46	23	67	1.5	4.5	8.0	6.0	2.0	15.4	6.3	4.7	1.6	13.5
7	220	3	660	4.16	9	5.94	4	417	1667	4.38	1.25	2.08	54	27	71	1.9	4.5	8.0	6.4	1.6	17.0	6.0	4.8	1.2	14.7
8	220	3	660	4.16	9	5.94	4	417	1667	4.38	1.25	2.08	63	31	75	2.4	4.5	8.0	6.9	1.2	18.2	5.7	4.9	0.8	15.5
9	220	3	660	4.16	8	5.28	4	417	1667	4.38	1.25	2.08	71	36	80	2.8	4.5	7.4	7.3	0.0	18.2	5.0	5.0	0.0	15.5
10	220	3	660	4.16	8	5.28	4	417	1667	4.38	1.00	1.67	80	40	84	3.4	4.5	6.9	7.9	-0.9	17.3	4.5	5.1	-0.6	14.9
11	220	4	880	5.55	8	7.04	4	417	1667	4.38	1.00	1.67	90	45	90	4.0	4.5	8.7	8.5	0.2	17.5	5.3	5.2	0.1	15.0
12	220	4	880	5.55	8	7.04	4	417	1667	4.38	1.00	1.67	100	50	95	4.7	4.5	8.7	9.2	-0.5	16.9	5.1	5.4	-0.3	14.7
13	220	4	880	5.55	7	6.16	4	417	1667	4.38	1.00	1.67	110	55	101	5.5	4.5	7.8	10.0	-2.2	14.7	4.4	5.6	-1.2	13.5
14	220	5	1100	6.94	7	7.7	4	417	1667	4.38	1.00	1.67	121	61	107	6.5	4.5	9.4	11.0	-1.6	13.2	5.0	5.8	-0.8	12.7
15	220	5	1100	6.94	7	7.7	4	417	1667	4.38	1.00	1.67	132	66	113	7.5	4.5	9.4	12.0	-2.6	10.5	4.7	6.1	-1.3	11.4
16	220	5	1100	6.94	7	7.7	4	417	1667	4.38	1.00	1.67	144	72	120	8.6	4.5	9.4	13.1	-3.7	6.8	4.5	6.3	-1.8	9.6
17	220	5	1100	6.94	6	6.6	4	417	1667	4.38	1.00	1.67	155	77	127	9.8	4.5	8.3	14.3	-6.1	0.7	3.8	6.6	-2.8	6.8
18	220	5	1100	6.94	6	6.6	4	417	1667	4.38	1.00	1.67	166	83	135	11.2	4.5	8.3	15.7	-7.4	-6.7	3.6	6.8	-3.2	3.5
19	220	5	1100	6.94	6	6.6	4	417	1667	4.38	1.00	1.67	178	89	143	12.7	4.5	8.3	17.2	-8.9	-15.6	3.4	7.1	-3.7	-0.2
20	220	5	1100	6.94	6	6.6	4	417	1667	4.38	1.00	1.67	189	94	151	14.3	4.5	8.3	18.8	-10.5	-26.1	3.3	7.4	-4.2	-4.3
21	220	5	1100	6.94	5	5.5	4	417	1667	4.38	1.00	1.67	200	100	160	16.1	4.5	7.2	20.6	-13.4	-39.5	2.7	7.7	-5.0	-9.4
22	220	5	1100	6.94	5	5.5	4	417	1667	4.38	1.00	1.67	212	106	170	18.0	4.5	7.2	22.5	-15.3	-54.8	2.6	8.1	-5.5	-14.9
23	220	5	1100	6.94	5	5.5	4	417	1667	4.38	1.00	1.67	223	111	180	20.1	4.5	7.2	24.6	-17.4	-72.2	2.4	8.4	-6.0	-20.8
24	220	5	1100	6.94	5	5.5	4	417	1667	4.38	1.00	1.67	234	117	191	22.4	4.5	7.2	26.9	-19.7	-91.9	2.3	8.7	-6.4	-27.2
25	220	5	1100	6.94	4	4.4	4	417	1667	4.38	1.00	1.67	246	123	202	24.9	4.5	6.1	29.4	-23.3	-115.2	1.9	9.1	-7.2	-34.5
26	220	5	1100	6.94	4	4.4	4	417	1667	4.38	1.00	1.67	257	128	215	27.6	4.5	6.1	32.1	-26.0	-141.2	1.8	9.5	-7.7	-42.1
27	220	5	1100	6.94	4	4.4	4	417	1667	4.38	1.00	1.67	268	134	227	30.5	4.5	6.1	35.0	-28.9	-170.2	1.7	9.8	-8.1	-50.3
28	220	5	1100	6.94	4	4.4	4	417	1667	4.38	1.00	1.67	279	140	241	33.7	4.5	6.1	38.2	-32.1	-202.3	1.6	10.2	-8.6	-58.9
29	220	5	1100	6.94	4	4.4	4	417	1667	4.38	1.00	1.67	291	145	256	37.2	4.5	6.1	41.7	-35.6	-237.9	1.5	10.6	-9.1	-68.0
30	220	5	1100	6.94	4	4.4	4	417	1667	4.38	1.00	1.67	302	151	271	40.9	4.5	6.1	45.4	-39.4	-277.2	1.5	11.0	-9.6	-77.5
Totals		123 Units	27 GW	171 Twh p.a.		170 $B		12500 Units	50 GW	131 Twh p.a.		59 $B				371 $B	135 $B	229 $B	506 $B	-277 $B	-277 $B	126 $B	203 $B	-78 $B	-78 $B

30 Year 100% Renewable Energy Implementation Plan - $120/tonne Coal Starting Price

General Findings:

1. Cost of 100% renewable energy rollout	**$229 B**
2. Savings from reduced coal usage	**$1.025 B**
3. Profit of plan over 30 years	**$797 B**

Assumptions	Australia's current electricity usage	257 TWh p.a.		
	Plus 30% population growth	334 TWh p.a.		
	Minus 10% efficiency gain	301 TWh p.a.		
	Starting cost of solar thermal	10.5 $M/MW		
	Starting cost of wind power	2.2 $M/MW		
			CST Part	Wind part
Key Results	Power capacity rating	77 GW	27	50
	Power capacity provided	302 TWh p.a.	171	131
	Cost of implementation	229 $B	170	59
	Average capital cost over 30 years	3.0 $M/MW		
			CST Part	Wind part
Comparison with BZE 10 Year plan	Power capacity rating	93 GW	43	50
	Power capacity provided	325 TWh p.a.	195	131
	Cost of implementation	261 $B	190	71
	Average capital cost over 10 years	2.8 $M/MW		

General notes

1. The BZE 10 year plan assumes greater future efficiency gains, but also costs electrical energy to service conversion of gas heating and transportation to electricity.
2. The BZE 10 year plan also provides $90B for transmission line upgrades and $17B for additional off-grid and backup bio-mass systems.

   My book's full 30 year plan funds $255B which therefore covers these latter items. Transmission line upgrades also occur in the BAU case, estimated by BZE at $2B p.a.
3. Total costs therefore depend on overall plan assumptions. This plan is based on current technology. Technology advances over the 30 years will result in a better outcome.

Concentrating solar thermal assumptions

1. As per BZE plan, each 220 MW unit typically creates 1388 GWh p.a. Starting price is $10.5M/MWe, reducing roughly in line with the European Commission's 2008 analysis.
2. This analysis assumes slower CST price reduction over time than the BZE plan, but the longer 30 year timeframe takes greater advantage of reduced prices.

Wind tower assumptions

1. As per BZE plan, assumes 30% efficiency per tower, and similar expected price reductions over first 10 years.

Coal savings assumption

1. 1 Kg coal generates 2kWh electricity, 1MT coal generates 2TWh electricity
2. As estimated by BZE new coal plant capital investment under BAU scenario is $135B over 30 years.
3. This analysis includes capital costs of renewable energy, fuel costs of coal, and capital costs of coal, being the three largest factors.
4. Black coal is the dominant current energy source. Costs vary over time. This analysis assumes start at either $50 or $120/ton and increasing 6% p.a. Coal prices follow oil. These coal prices should be seen as a combined price for a black coal, brown coal, gas fossil fuel mix. Brown coal is cheaper and gas is more expensive, at least historically. Coal prices per tonne and per ton are similar. In reality, fuel cost savings would be offset one year from the analysis shown below.
5. Other operating and maintenance costs of coal or renewable energy are similar and covered in more detail in the BZE analysis and the attached LEC analysis.

	Concentrating Solar Thermal Capital Costs (Annual)						Wind Tower Capital Costs (Annual)						Coal Usage Saved (Annual)				Saved New Coal Plants Capital $B	Total Costs/Savings (Annual)		Plan Cost (savings are negative)		Future Discounting at 5% p.a.			
Year	MWe/unit	Units p.a.	MWe/p.a.	New TWh p.a capacity	$M/MWe	$B	MW/unit	Units p.a.	MW/p.a.	New TWh pa capacity	$M/MW	$B	Coal Twh p.a. Replaced	Coal MT Replaced	Coal Cost/Ton $	Coal Cost saved p.a. $B		Total Costs Renewable Energy $B	Total Coal Savings $B	Annual Net Plan Cost $B	Cumulative Plan Cost $B	Annual Plan Cost $B	Total Coal Savings $B	Annual Net Plan Cost $B	Cumulative Plan Cost $B
1	220	0	0	0.00	10.5	2.31	4	417	1667	4.38	2.2	3.67	4	2	120	0.3	4.5	6.0	4.8	1.2	1.2	6.0	4.8	1.2	1.2
2	220	2	440	2.78	10.5	2.31	4	417	1667	4.38	1.9	3.17	12	6	127	0.7	4.5	5.5	5.2	0.2	1.5	5.2	5.0	0.2	1.4
3	220	3	660	4.16	10.5	6.93	4	417	1667	4.38	1.9	3.17	20	10	135	1.4	4.5	10.1	5.9	4.2	5.7	9.2	5.3	3.8	5.3
4	220	3	660	4.16	9	5.94	4	417	1667	4.38	1.65	2.75	29	14	143	2.0	4.5	8.7	6.5	2.1	7.8	7.5	5.7	1.9	7.1
5	220	3	660	4.16	9	5.94	4	417	1667	4.38	1.65	2.75	37	19	151	2.8	4.5	8.7	7.3	1.4	9.2	7.1	6.0	1.1	8.3
6	220	3	660	4.16	9	5.94	4	417	1667	4.38	1.25	2.08	46	23	161	3.7	4.5	8.0	8.2	-0.1	9.1	6.3	6.4	-0.1	8.2
7	220	3	660	4.16	9	5.94	4	417	1667	4.38	1.25	2.08	54	27	170	4.6	4.5	8.0	9.1	-1.1	8.0	6.0	6.8	-0.8	7.3
8	220	3	660	4.16	9	5.94	4	417	1667	4.38	1.25	2.08	63	31	180	5.7	4.5	8.0	10.2	-2.1	5.8	5.7	7.2	-1.5	5.8
9	220	3	660	4.16	8	5.28	4	417	1667	4.38	1.25	2.08	71	36	191	6.8	4.5	7.4	11.3	-4.0	1.9	5.0	7.7	-2.7	3.1
10	220	3	660	4.16	8	5.28	4	417	1667	4.38	1.00	1.67	80	40	203	8.1	4.5	6.9	12.6	-5.7	-3.8	4.5	8.1	-3.6	-0.5
11	220	4	880	5.55	8	7.04	4	417	1667	4.38	1.00	1.67	90	45	215	9.7	4.5	8.7	14.2	-5.4	-9.2	5.3	8.7	-3.3	-3.8
12	220	4	880	5.55	8	7.04	4	417	1667	4.38	1.00	1.67	100	50	228	11.4	4.5	8.7	15.9	-7.2	-16.4	5.1	9.3	-4.2	-8.0
13	220	4	880	5.55	7	6.16	4	417	1667	4.38	1.00	1.67	110	55	241	13.2	4.5	7.8	17.7	-9.9	-26.3	4.4	9.9	-5.5	-13.5
14	220	5	1100	6.94	7	7.7	4	417	1667	4.38	1.00	1.67	121	61	256	15.5	4.5	9.4	20.0	-10.6	-36.9	5.0	10.6	-5.6	-19.2
15	220	5	1100	6.94	7	7.7	4	417	1667	4.38	1.00	1.67	132	66	271	18.0	4.5	9.4	22.5	-13.1	-50.0	4.7	11.3	-6.6	-25.8
16	220	5	1100	6.94	7	7.7	4	417	1667	4.38	1.00	1.67	144	72	288	20.7	4.5	9.4	25.2	-15.8	-65.8	4.5	12.1	-7.6	-33.4
17	220	5	1100	6.94	6	6.6	4	417	1667	4.38	1.00	1.67	155	77	305	23.6	4.5	8.3	28.1	-19.9	-85.6	3.8	12.9	-9.1	-42.5
18	220	5	1100	6.94	6	6.6	4	417	1667	4.38	1.00	1.67	166	83	323	26.9	4.5	8.3	31.4	-23.1	-108.7	3.6	13.7	-10.1	-52.6
19	220	5	1100	6.94	6	6.6	4	417	1667	4.38	1.00	1.67	178	89	343	30.4	4.5	8.3	34.9	-26.6	-135.4	3.4	14.5	-11.1	-63.6
20	220	5	1100	6.94	6	6.6	4	417	1667	4.38	1.00	1.67	189	94	363	34.3	4.5	8.3	38.8	-30.5	-165.9	3.3	15.4	-12.1	-75.7
21	220	5	1100	6.94	5	5.5	4	417	1667	4.38	1.00	1.67	200	100	385	38.5	4.5	7.2	43.0	-35.9	-201.8	2.7	16.2	-13.5	-89.2
22	220	5	1100	6.94	5	5.5	4	417	1667	4.38	1.00	1.67	212	106	408	43.2	4.5	7.2	47.7	-40.5	-242.3	2.6	17.1	-14.5	-103.8
23	220	5	1100	6.94	5	5.5	4	417	1667	4.38	1.00	1.67	223	111	432	48.2	4.5	7.2	52.7	-45.5	-287.8	2.4	18.0	-15.6	-119.3
24	220	5	1100	6.94	5	5.5	4	417	1667	4.38	1.00	1.67	234	117	458	53.7	4.5	7.2	58.2	-51.0	-338.8	2.3	18.9	-16.6	-135.9
25	220	5	1100	6.94	4	4.4	4	417	1667	4.38	1.00	1.67	246	123	486	59.6	4.5	6.1	64.1	-58.1	-396.9	1.9	19.9	-18.0	-153.9
26	220	5	1100	6.94	4	4.4	4	417	1667	4.38	1.00	1.67	257	128	515	66.1	4.5	6.1	70.6	-64.6	-461.5	1.8	20.9	-19.1	-173.0
27	220	5	1100	6.94	4	4.4	4	417	1667	4.38	1.00	1.67	268	134	546	73.2	4.5	6.1	77.7	-71.6	-533.1	1.7	21.9	-20.1	-193.2
28	220	5	1100	6.94	4	4.4	4	417	1667	4.38	1.00	1.67	279	140	579	80.9	4.5	6.1	85.4	-79.3	-612.4	1.6	22.9	-21.2	-214.4
29	220	5	1100	6.94	4	4.4	4	417	1667	4.38	1.00	1.67	291	145	613	89.2	4.5	6.1	93.7	-87.6	-700.0	1.5	23.9	-22.4	-236.7
30	220	5	1100	6.94	4	4.4	4	417	1667	4.38	1.00	1.67	302	151	650	98.2	4.5	6.1	102.7	-96.7	-796.7	1.5	25.0	-23.5	-260.2
Totals		**123 Units**	**27 GW**	**171 Twh p.a.**		**170 $B**		**12500 Units**	**50 GW**	**131 Twh p.a.**		**59 $B**				**890 $B**	**135 $B**	**229 $B**	**1025 $B**	**-797 $B**	**-797 $B**	**126 $B**	**386 $B**	**-260 $B**	**-260 $B**

Additional Impact of Carbon Price						
Year	Carbon price $/Tonne	Cents / KWh	Total TwH Used	Coal Twh Used	Carbon Price Total $B	Carbon Price Future Discounted $B
1	25	3	257	253	6.3	6.3
2	27	3	258	247	6.6	6.3
3	29	3	260	240	6.9	6.2
4	31	3	261	233	7.1	6.2
5	33	3	263	226	7.4	6.1
6	35	4	264	218	7.7	6.0
7	38	4	266	211	7.9	5.9
8	40	4	267	204	8.2	5.8
9	43	4	269	197	8.5	5.7
10	46	5	270	190	8.7	5.6
11	49	5	271	182	8.9	5.5
12	53	5	273	173	9.1	5.3
13	56	6	274	165	9.3	5.2
14	60	6	276	155	9.3	5.0
15	64	6	278	145	9.4	4.7
16	69	7	279	135	9.3	4.5
17	74	7	281	126	9.3	4.2
18	79	8	282	116	9.1	4.0
19	84	8	284	106	9.0	3.7
20	90	9	285	96	8.7	3.4
21	97	10	287	87	8.4	3.2
22	104	10	288	77	8.0	2.9
23	111	11	290	67	7.4	2.5
24	119	12	292	57	6.8	2.2
25	127	13	293	48	6.0	1.9
26	136	14	295	38	5.1	1.5
27	145	15	296	28	4.1	1.2
28	155	16	298	19	2.9	0.8
29	166	17	300	9	1.5	0.4
30	178	18	301	0	0.0	0.0
					217 $B	**122 $B**

Note to the reader

The previous letter updates this book and explains how the cost of implementing 100% renewable energy, as well as carrying out the remainder of a comprehensive climate change action plan that is described within this book, will be cheaper than staying with coal.

Despite the doubts that were expressed, a subsequent newspaper article titled "NSW's Great Big Coal Subsidy Scandal", suggested that the comparitive costings for coal of $50/tonne (subsidised) or $120/tonne or more (unsubsidised) were actually spot on.

The cost of implementing solar energy is now, in fact, close to or even less than the cost of implementing new coal fired power stations. We should not implement any more of these. The timeframe required to implement 100% renewable energy is long, but the savings and other environmental benefits from starting rapidly now, rather than delaying, will therefore be large.

The analysis spreadsheets attached to this previous letter show a detailed plan for renewable energy implementation over 30 years, along with the estimate of coal savings. They can be sent to you by contacting the author at the email address provided at the back of this book.

These new costings are based on solar thermal systems capable of meeting peak and baseload energy demands, using energy storage. These costings replace the renewable energy costing provided within the book, which yielded a similar total cost, but were based more upon solar photovoltiac (PV) systems without any energy storage. Solar PV can still play a part and can be used to reduce costs. The costing used within the book, of $1.50/watt, has now become a reality and is being surpassed with reports of costs for commercial solar PV installations now dropping as low as $1.00/watt.

The equivalent analysis of costs and savings for the USA can also be obtained by contacting the author at the email address provided at the back of this book.

There are further update letters provided at the back of the book.

The first of these is very important. It is titled "The Potential Climate Policy Impasse, How Current and Future Generations Should Pay for Renewable Energy". The previous letter shows the massive profits of carrying out a 100% renewable energy plan. It also shows a very small upfront investment cost of between $10B and $20B for Australia in the first 10 years before the massive profits start to kick in. This next letter explains how these small upfront costs can be shared between current and future generations. The upfront costs can also be met by borrowing so that the current generation need not pay anything.

The economic benefits to future generations are massive. The current generation need not be held fully responsible for meeting these startup costs. The current generation is however fully responsible for starting the plan.

The second of the letters is titled "The Greens Urgently Need to Become Australia's Sound Economic Managers". It suggests to the Australian Greens that the range of mechanisms that they seek to negotiate, such as high carbon prices and feed-in tariffs, will be seen as too expensive, unless a specific implementation plan is constructed such as the one provided in this book, so that the costs and benefits can be properly understood and explained.

The body of this book was written in 2010 in the lead up to the 2010 Australian federal election. It therefore analysed the Australian Government's proposed emissions trading scheme called the "Carbon Pollution Reduction Scheme (CPRS). Following that election and the announcement of the new Australian Government's revised carbon pricing policy, the "Clean Energy Future", a range of comparitive analysis material of this new carbon pricing policy with this book's proposed policy was developed at the book's web site. That web site material is no longer provided online, but is now provided as the last appendix of this book. It contains:

- a summary of this book's proposed climate policy for Australia and why it is more effective than carbon pricing
- a description of why Australia's planned carbon pricing approach is unlikely to be effective
- a summary of why a more effective policy need not cost us anything in the short term, and can generate massive savings over the medium to long term
- a call for petition for 300% renewable energy for Australia, leading to green energy export, at the same cost as the funding proposed to be sent overseas as part of carbon pricing and emissions trading, but with far greater benefits
- a high level design and costing for a world green energy system for 2080, with Australia exporting 10% of the world's green energy needs, and generating annual income many times higher than our current annual GDP

A recent article available on the web titled "A Tipping Point for Fossil Fuels" summarises what the International Energy Agency has to say about the costs and benefits of aggressive action. It discusses the need for large scale urgent action to begin immediately and the risks to the world of not doing so. It suggests a cost of $20 Trillion for renewable energy for the world, matching very closely the cost for a world energy system design provided at the back of this book, and agrees this is much cheaper than staying with fossil fuels.

If you prefer to jump to the book's final conclusions, then go to the back of the book and read the summary material from the book's web site. If not, grab a drink and settle into reading about the journey.

Preface

I became interested, and now involved, in the climate change debate after reading Australian skeptic Ian Plimer's book, "*Heaven + Earth - Climate Change: The Missing Science*" (1). It is a very good book. I was fascinated to read its well-presented and skeptical views about how we could be so misguided on the climate change issue. It prompted me to want to know the most likely truth.

The science

My research project thus began in February of 2010. Over the following months I spent much time sifting through the scientific arguments and the opposing skeptical arguments. This more detailed examination showed the skeptical arguments to be less persuading and the mainstream science to be well thought through. I thus gradually came to the conclusion that the mainstream scientists are probably correct and the skeptics are probably wrong.

I also observed that there are few good texts explaining specifically why climate change skeptics are probably wrong, and so this book began. The mainstream scientific views and views of the skeptics are discussed in Part A of this book.

The currently proposed policies

I then moved on to look at the climate change policy environment, in terms of what the world is doing to address the significant risks of climate change, and in particular what my own country Australia is doing. I came to the conclusion that neither what the world nor Australia are proposing are anywhere near adequate to address the significant risks of climate change. The currently proposed policies in countries such as Australia and the USA, and suggestions for better, more effective policies, are explored in Part B of this book.

The current Australian Government originally proposed a climate change policy based on an emissions trading scheme. Following the Copenhagen Accord, Australia offered commitments of 5% emissions reductions by 2020, or up to 25% reductions in the same timeframe if other world governments did the same. The proposed emissions trading scheme could be scaled to meet any emission reduction targets. A number of emission reduction scenarios were analysed and the highest level scenarios targeted aggressive 80-90% emission reductions by 2050.

The highest level emission reduction scenario matched the level of emission reductions that scientists believe might give us a 50% chance of limiting warming to 2 degrees. 2 degrees of warming is the target set at Copenhagen and is the level at which it is thought the world would avoid "dangerous" levels of climate change. Achieving this target however requires international action. There was much concern in Australia about the emissions trading proposal. It has cost uncertainties and could result in job losses.

As a result of opposition to the policy, an inability to get the legislation through the Australian Senate, and uncertainty regarding the world's response to climate change, the Australian Government has delayed its proposed implementation until 2013. Since the 2010 Australian general election, other proposals such as a carbon tax are now also being considered.

The Australian Opposition party has proposed an alternative low cost direct action policy. It claims to match the current Government's base proposal of a 5% emission reduction target by 2020, but at about one tenth of the cost. The bulk of the proposed emission savings are based upon attempting to cheaply bury carbon in the soil rather than actually reducing emissions. There is debate whether this proposed "soil abatement" approach would really work. The policy does not discuss any strategies to scale to higher emission reductions that would meet Copenhagen Accord targets, or any targets beyond 2020. Unfortunately, because of the concerns over emissions trading, and a level of skepticism about climate change in general, Australia's public might prefer this low cost and unproven approach to achieving only modest emission reductions.

Better policies?

Both of these original Australian policy proposals have some troubling weaknesses. Of the two policies, the emissions trading proposal is the better attempt, and describes an ability to scale to stronger emission reduction targets. Some economists describe it as the most economically "efficient" method. However there is also skepticism whether emissions trading is really an effective method for climate change mitigation at all. Some of its problems, as well as carbon pricing in general, are described in this book.

In order to avoid some of the profiteering, scams and rorting that have resulted from free handouts of emission permits in the European emissions trading scheme, emission permits in Australia will, for the most part, be auctioned, rather than handed out for free.

The expected pool of funds from these auctions is substantial, starting at around $13B per year in initial years. The opposition party presents this figure as the cost of the emissions trading proposal. The public might generally have a perception that these funds will be used to fight climate change, but this is not the case. The theory of emissions trading discourages such funds being used that way, and suggests they should mainly be used for compensating household subsidies or a relatively small amount of climate change R&D.

The large pool of funds gained from the permit auctions could provide the most substantial vehicle for Australia to fight climate change, rather than the emissions trading process itself, which works more around the edges to make only gradual and small emissions cuts each year. None of the large pool of permit funds actually need be spent on climate change in order for the emissions trading process to work, and emission reduction targets and caps to be met, but so much more could be achieved if these funds were effectively used. Similar arguments can be made for how revenue from an alternative carbon tax proposal might be used.

This book puts forward a more aggressive climate change policy approach that Australia, and other countries of the world, could adopt in order to not only match the Copenhagen Accord targets, but better them.

This book proposes that Australia, and other countries, could implement 100% renewable energy and also be carbon neutral by 2040, if we tried, but only if all potential climate change mitigation revenues are most efficiently and strategically spent. This level of aggressive action could give us a better than 50% chance of avoiding "dangerous" climate change. Surely, this is what we want.

Although the policy discussion in this book focuses on Australia, the book also discusses what could be done by the world to address climate change, particularly for the five largest emitting nations, which account for over 50% of the world's emissions. Australia is a small proportion of the overall world population, and carbon emissions, but we do have an ability to take proper action and also to have considerable influence on the world as a whole. This partly comes about because of our strong ties with the USA and China. This book scales a similar policy proposed for Australia to the larger size of the USA. It also makes suggestions for the basis of the next world climate change accord.

The costs vs the risks

The cost of work proposed in the policy documents and in this book may look very large. However, they are tiny when viewed at a proportion of GDP (less than 1%), per capital or per household (a few hundred dollars per year).

The costs are insignificant when compared to the enormous risks involved in "dangerous" levels of climate change, both in terms of damage to our economies and environment, risks to food supply, extinction of species and loss of human life on potentially very large scales. A detailed discussion of the enormous risks of climate change is not provided in this book as I want the book to be easily read by anyone. I have left references to relevant risk assessment documents at the back of this book.

From what I have seen, the quality and quantity of the scientific work being done around the world is very good. There is a huge amount of advice being given to our Governments about the science of climate change, its likely impacts, and the level at which we need to respond with mitigation strategies.

However, world governments are only partly responding. There are many arm-chair scientists, a bit like me, who seek to question whether the science is correct or not. It is hugely concerning that we are not, as a whole, listening to the scientific advice, and acting on it in a responsible manner. This could unfortunately bring about the worst forecasted effects of climate change. The general public is for the most part not aware of the extent of the risks, and they are actually beyond anyone's full comprehension. The last time science had such a hard time being heard was at the time of Galileo. However some recent surveys suggest the majority of public believe climate change is most likely human caused and want something done about it.

Based on the size of the clearly presented risks, we should be over-estimating the amount of climate change action that we need to take. At present we seem intent on under-estimating it.

The policy alternative that is put forward in this book for Australia will cost $13B p.a. How much is $13B p.a. for Australia? It is about 1% of GDP. It is equal to the Government's proposed price for emission permits in early years, but is more than we are proposing to spend on climate change mitigation. It is less than our Defence budget of around $17B p.a. Climate change is the bigger risk. It is similar in size to the new proposed mining tax. It is more than we spend on roads, at around $9B p.a. It is less than we spend on health. It is more than we spend on building new electricity plants at present, at around $2.8B p.a. As a result of Australia's current resources boom, which is based mainly on fossil fuels, it is less than one year of Australia's expected GDP growth.

The same policy for the USA will cost $172B p.a.

If we chose to be even more aggressive, we would double this book's proposed annual expenditure to 2% of GDP p.a., and aim to achieve 100% renewable energy and 100% emissions reductions in 15 years, not 30. This would be in the aim of reducing the already expected economic losses due to climate change towards the end of this century, and into the next, and also to help avoid the risk of even more dangerous effects of climate change.

My hope is that the ideas presented in this book will help raise awareness of what the science is saying, the risks we are facing, and of the policy options that we are being offered.

Finally

So in summary, what is this book? I would summarise it as follows. Part A is a summary of mainstream scientific and also skeptical views. It analyses the skeptical view and explains why they are probably wrong. Part B analyses currently proposed policies in Australia and then proposes a very different, more aggressive large-scale direct-action policy. In my view, emissions trading, and other forms of carbon pricing such as a carbon tax, will tend to just work around the edges of the problem in a slow and uncoordinated and probably ineffective way. The book is a proposal that we embark on a large scale programme of coordinated and urgent climate change mitigation.

There are few texts that explain both the advantages and also the disadvantages of carbon pricing approaches, such as emissions trading or a carbon tax. This book does that.

I recommend that you read this book if:

- You are a voter and want to better understand the scientific and policy climate change debate.
- You are a policy maker and want to know how to improve your policies or how you are describing your policies.
- You are a concerned citizen of the world and want to know how we might have a better than 50% chance of avoiding "dangerous" levels of climate change.

A strategy is put forward for achieving 100% renewable energy and being carbon neutral in 30 years, for Australia. A similar policy might also work for other nations. In an attachment to this book an almost identical plan is also provided for the USA to also implement 100% renewable energy and be carbon neutral in 30 years. Such a call is not unusual. For example the work

"ZeroCarbonBritain" (67), written and reviewed by a huge team of climate change experts, calls for Britain to achieve the same result in twenty years. The requirement for the world to spend 1% of GDP in order to fight climate change is also documented in other texts.

In Part A of the book, we build a simple climate change model. It's not as good as our scientist's ones. In Part B, we build a simple climate change mitigation policy. It is better than our politician's ones. I am however not a scientist, and also not an economist, and so the book needs to be read in that light.

As a bit of fun, and as a reference to the previously mentioned book, I was toying with an alternative title for the book of "*Science + Politics - Climate Change Mitigation: The Missing Policies*". However I don't want the title of this book to appear that negative. This book differs from other texts on climate change in that it spend less time describing the problem of climate change and more time describing, costing and planning some aspirational solutions.

I've tried to keep the book relatively simple. I hope you enjoy it.

The Author

Postscript - on the Australian 2010 election and a carbon tax. This book was originally and mainly written prior to the Australian 2010 federal election. As a result of the Australian 2010 federal election the incumbent Government only just managed to hold onto office, based on a tenuous level of support from the Greens party and a small group of independents. Climate change was not really a major area of election campaign debate, possibly because neither major political party really had a strong climate change policy to sell.

Where this book discusses the next election, now read 2013, as Australian federal elections are normally held every three years. This coincides with the Australian Government's previously suggested date for the start of an emissions trading scheme. A multi-party Climate Change Committee has been set up to analyse a range of different policy options during this time.

As a simpler alternative to an emissions trading scheme, or as an intermediate step, a carbon tax is now proposed. This book describes why a tax alone will not necessarily cause industry or Government to carry out an adequate level of action to mitigate the risks of climate change.

Much of the discussion of emissions trading contained in this book is relevant and necessary reading as a background to considering the option of a carbon tax. As a carbon tax is now also firmly on the potential policy agenda for Australia, and possibly also the USA, a new chapter discussing a carbon tax alternative has been added to this book.

A carbon tax does not guarantee a level of action. If the price of carbon is too low, it will simply be a cost inconvenience for industry and consumers. If new prices are just accepted, with the price of carbon too low and subsidies too high, an inadequate level of action will be taken. Also, because of a number of "non-price barriers" to action, industry will not necessarily act as government expects it to. These other barriers to action are described within this book.

The principle of a carbon tax is to cause pricing-pain and thus encourage action. However the amount of action taken will generally be less than the amount of pain caused. From that perspective, it isn't clear whether a carbon tax is an efficient, effective or reliable method to achieve climate change mitigation as some economists claim.

Proponents of a carbon pricing approach, particularly from industry, argue that it is better than a direct action approach because it avoids Governments having to "pick winners". If we are going to act, one way or another someone needs to start picking and implementing some "winners".

The large-scale direct-action policy provided in this book describes how industry and government could adopt a flexible strategy and approach in a 30 year plan with much more certainty of solving the problem of climate change.

Table of Contents

PART A – THE SCIENCE

A summary of the science

As a summary, examination of the science and evidence suggests the following statements are all probably true:

1. The earth's temperature is rising, and has risen by about 0.7 degree over the past century;
2. CO2 levels in the atmosphere are rising and will reach double pre-industrial levels in a few decades;
3. Carbon dating analysis provides evidence that the additional CO2 in our atmosphere is human caused, from burning very old fossil fuels. Also, the nature of the warming, including warming at night, at the poles and the troposphere (lower atmosphere) has the signature of an increased greenhouse effect;
4. Modeling of the effects of the sun, CO2 and many other climate contributors suggests that CO2 is probably the major cause of temperature rises over the past half century, and increasing CO2 levels will result in even more significant climate and weather changes over the next decades;
5. Climate models, and analysis of CO2 levels in the past, suggest a doubling of CO2 in our atmosphere will result in a rise in temperature of about 3 degrees. If we don't act aggressively, this level will be reached in only a few decades. This is about 4 times the rise in temperature seen so far over the past century, and this will substantially and quickly change our climate, far beyond levels we have seen so far; and
6. Temperature rises could go higher than 3 degrees, with potentially very severe consequences.

Climate change skeptics will generally argue one or more of the following:

A. Temperatures aren't rising;

B. CO2 levels aren't rising;

C. Humans aren't causing CO2 levels to rise;

D. Increasing CO2 won't cause temperatures to rise; and/or

E. Rising temperatures won't matter.

Some skeptics argue all of the above. However, and unfortunately, closer examination of the science and evidence suggests that none of these skeptical arguments are likely to be true.

The following chapters describe more detail of the scientific view. This is followed by a more detailed look at suggestions made by the skeptics, and why those are probably wrong.

Reasons to believe humans are causing climate change

Temperatures are rising

The graph below shows the rise in global temperatures since the start of industrialisation. It is characterised by a numbers of rises and falls, and an overall rising trend. There is little debate about this data, which has been gathered using reasonably accurate thermometers.

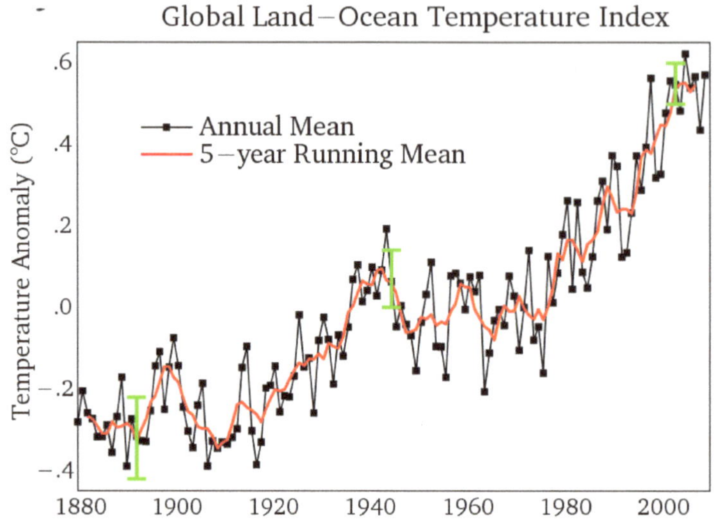

(2)

You will notice that similar rates of temperature rise in recent decades also occurred earlier in the last century. Computer models of the earth climate suggest that in the first half of the last century the temperature rises were largely due to increased solar activity. The sun's power is always slightly varying. It is thought the temperature rises in the latter half of the century were mostly caused by rising carbon dioxide (CO2) levels in the atmosphere. Carbon dioxide is a "greenhouse" gas. Both of these effects are discussed further in later chapters.

There has been some debate whether the earth has continued to warm over the past 10 years. The graph below of temperatures since 1996 portray a continuing overall rising trend.

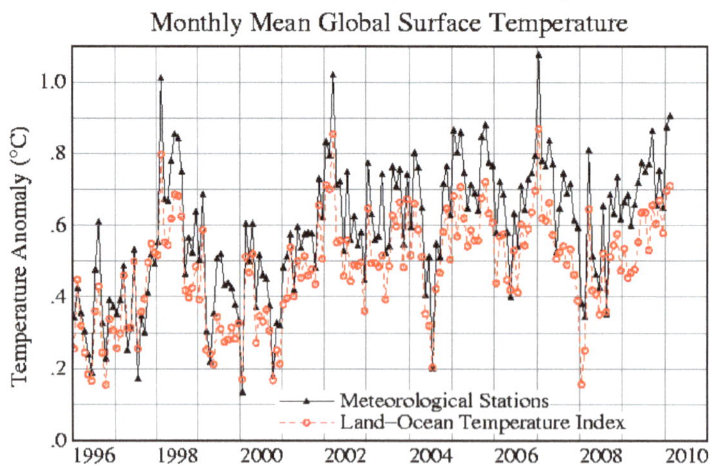

(11)

Mainstream scientists say that short term 5 to 10 year cyclic effects should not be considered in considering overall patterns, and that a 25 year view is required.

Atmospheric CO2 levels are rising

Modern measurements taken at Mauna Loa at Hawaii records show a clear and steady rise of CO2 from start of measurement in 1959 to now. This is the standard world measuring site which is compared and calibrated against many other CO2 measuring sites around the world.

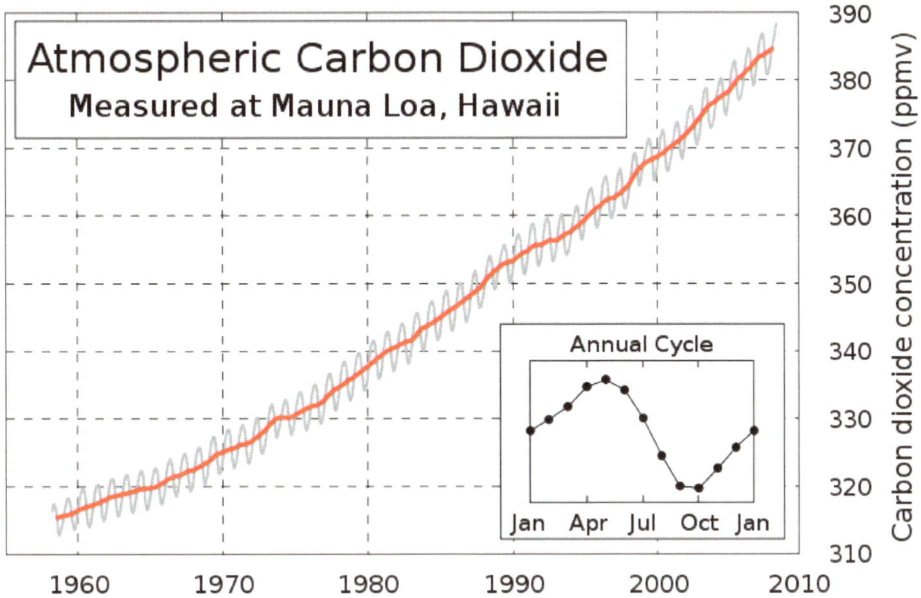

(3)

The data shows a steadily accelerating rise from 320 ppm (parts per million) to 390 ppm now. This easily extrapolates back to the estimated 285 ppm at the start of industrialization.

Analysis of CO2 levels in ice cores, drilled far down through very old ice packs, suggests that CO2 levels were reasonably constant at this pre-industrial level over past 1000 years. Skeptics often like to question whether this starting figure is accurate. Measurements of CO2 vary from location to location, but most measurements support the mainstream scientific view.

Are humans causing this CO2 increase?

The amount of carbon being put into the atmosphere by human activity is only a small part of the total carbon transfer cycle around the earth's atmosphere, oceans and "biosphere" (plants and animals).

However it is still large enough proportion to cause a probable building effect.

Carbon isotope analysis (C-14, C-13, C-12 – carbon atoms with different numbers of neutrons) suggests that the additional CO2 is of fossil fuel origin. This carbon dating techniques can show how old material is, based on the level of carbon 14, which is radioactive. It starts at known levels and gradually depletes over time according to its radioactive "half life". The carbon in fossil fuels is very old and includes virtually no carbon 14 left, and is also depleted in carbon 13. The changes in the type of carbon in our atmosphere are reported to match with an increased level of carbon from fossil fuels.

Note: the author found some references to this claim (e.g. 10), but did not find the source documents. There is still some debate whether it can be proven that the additional CO2 in the atmosphere is caused by humans. In practice, all the CO2 moves slowly around its normal, natural cycle.

CO_2 is a significant greenhouse gas - more of it is likely to warm the Earth

There is evidence to confirm that the recent warming of the earth is being caused by an increase in greenhouse gasses. Different causes of warming have different "fingerprints". Reasons to suspect an increasing greenhouse effect include increased warming at night time, warming at the poles, warming of the troposphere (lower part of the atmosphere) and cooling of the stratosphere (higher part of the atmosphere).

It isn't generally disputed that CO_2 contributes to the greenhouse effect and punches above its weight. CO_2 allows visible light through but absorbs infrared energy, and becomes warm. Other more plentiful gasses in the atmosphere such as nitrogen and oxygen do not do that. Other gasses such as methane do it even more so.

The total greenhouse effect currently warms the earth by about 19 degrees. Much of the greenhouse effect occurs in the upper atmosphere. The gases absorb energy and radiate heat both back to the earth and into space, working a bit like a blanket.

Although CO_2 is only 0.06% of the atmosphere, the contribution of CO_2 to the greenhouse effect is in the range is estimated to be 9-26% (4). CO_2 is the second largest contributor. Water vapour contributes the most – it is only 0.3% of total atmospheric mass (according to reference 4) but together with clouds contributes 66-85% (4 again). The effect of additional CO_2 is thought to occur mainly in the upper atmosphere where there is less contribution or feedback effects from water vapour.

Here are some of the important estimated figures.

- The total "warming force" of the total greenhouse effect is 171 watts/m2 (watts per square metre), including water vapour and clouds (5).
- One estimate of the warming force of all the CO_2 in the atmosphere is 23 watts/m2, 13% of the total greenhouse effect (5).
- The estimated warming force of additional green house gases since beginning of industrialisation is 2.45 watts/m2. The additional CO_2 contributes 1.67 watts/m2 or 68% of this total.
- The effect of additional CO_2 is estimated to rise to 3.7 watts/m2 at "2 X C02" (when CO_2 reaches double pre-industrial levels - 570 ppm).
- Other models suggest that, in a clear sky, CO_2 currently contributes up to 26% of the full 19 degrees or 20 watts/m2 greenhouse effect (6), which is approximately 5 degrees or 5 watts/m2.

The formula for the increase in CO_2's "warming force" in watts/m2 since industrialization is estimated as

$$5.35 \times \ln(CO_2\ \text{now} / CO_2\ \text{then})$$

Thus the rise in warming force is logarithmic, not linear, in that the same increase in warming force will occur with each doubling of the amount of CO_2.

Scientists differ on the level of temperature rise that will occur as a resulting of the additional warming force from additional CO_2. Feedback effects from increased evaporation, and resulting water vapour and clouds are difficult to model. Different models thus suggest different levels of "climate sensitivity", with a different factor to convert additional CO_2's warming force to temperature change generally somewhere between 0.2 to 1.0 as an average across the earth i.e.

$$dT = dE \times a$$

where dT is change in temperature, dE is additional warming force in watts/m2, and a is a factor that is generally thought to be somewhere between **0.2** and **1.0.**

There is much debate as to what the real factor is, and some suggest the climate is not sensitive at all (less than 0.2) and some suggest it could be more sensitive than this (greater than 1.0). It is this actual climate sensitivity that is going to either save us from the worst forecasts of temperature change, or not.

Using these two formulas it is easy to build a simple climate model spreadsheet that predicts the increase in earth's temperatures as CO2 levels rise, assuming different assumptions of climate sensitivity. If you do this, as the author has, you will quickly see what the scientists are warning us about. Assume CO2 levels double in the first half of this century, and double again by its end, as one worst case scenario that matches our current emissions activity.

Scientists have built much more complex computer models of how a range of different factors will warm or cool the earth. These models take into account factors such as rising CO2, changes in solar radiance, the cooling affect of aerosols in the atmosphere, feedback effects from water vapour and clouds, effects of ocean currents, amongst other factors.

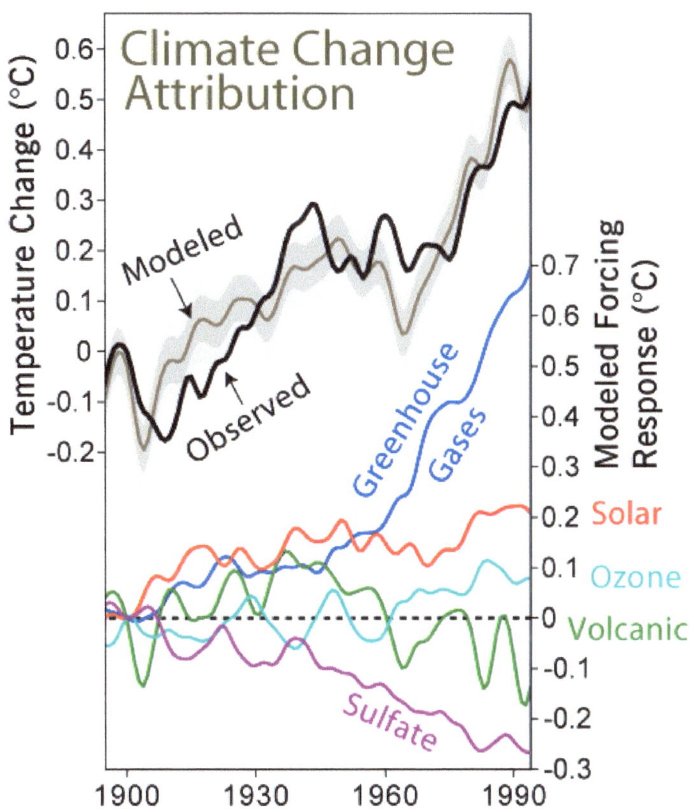

(7)

The modeling suggests that rising levels of CO2 is the main effect causing temperature rises in the second half of the last century.

Skeptics often argue that perhaps warming is caused by increased activity from the sun, not by increased CO2. Solar effects, shown as the red line in the diagram above, are thought by climate scientists to have been the larger factor in the first half of the last century. There is an interesting article on the Internet that says increased sunlight in the 1940's caused more glacier melting on the Matterhorn than that observed today (8). However solar effects do not vary significantly over short time periods under consideration, as suggested in the graph above. More discussion of this is provided in later chapters.

Recent temperature rises are unusual

Detailed analysis of many temperature records and proxies over the last 1000 years suggest that the abrupt rise over the last century is unusual, although not unprecedented.

The chart below shows a range of temperature measurements, using real thermometers and "proxy" measurements estimated from a variety of other data sources and locations.

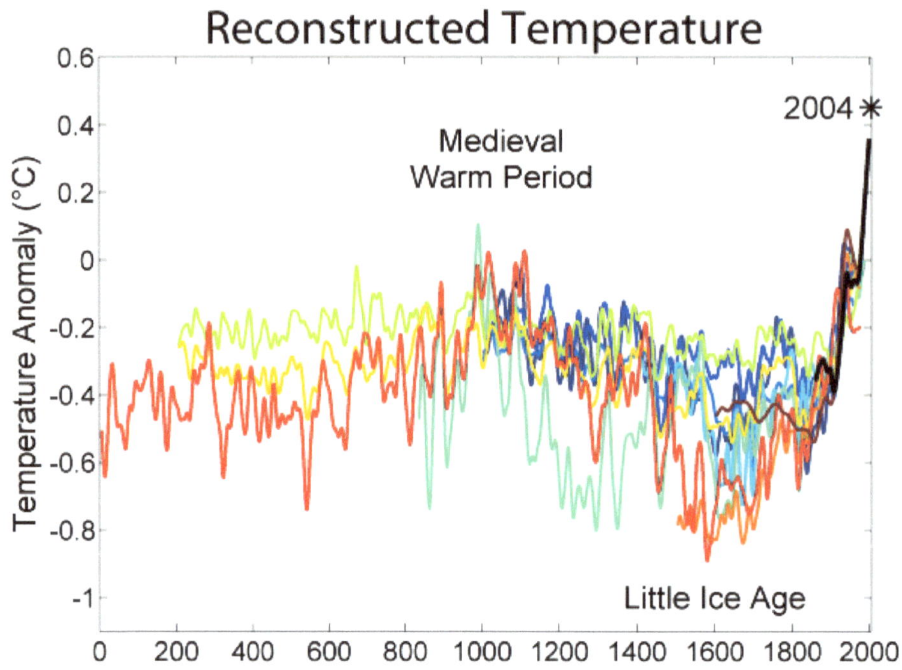

(9)

It is broadly accepted that temperatures were warm about 1000 years ago, in a period known as the "Medieval Climate Optimum". Also temperatures were cool in the 17th and 18th centuries, in a period known as the "Little Ice Age".

There is debate whether current temperatures are greater or less than "Medieval Climate Optimum" temperatures. The mainstream scientific view does not see this as an important consideration in the debate over what is likely to occur over the next century, where the forecast changes could be well in excess of 2 degrees. Cyclic rises and falls of temperature are however common over a variety of time periods. Some substantial changes over a longer time period are discussed later.

In contrast to the potentially large temperature rises forecast for the next century, temperature changes over the past 2000 years are thought to be generally only within only 1 degree.

Arguments from the skeptics and why they are probably wrong

It is reasonable to maintain some doubt about the climate change predictions, as climate scientists do not state their conclusions with 100% certainty.

This chapter discusses some of the arguments put forward by skeptics that substantial climate change will not occur, or that humans are not causing it. For reasons that are described, none of these counter arguments seem convincing when analysed in some detail. The explanation of the science was reasonably simple. Getting through all the main skeptical arguments takes a bit longer, as there are quite a few of them.

Clouds may reduce the warming effects of CO2

Climate scientists acknowledge there is limited understanding of the feedback effects on temperature from the combined effect of water vapour and clouds.

Warming from CO2 will result in additional water vapour and clouds. The feedback effect of water vapour is positive overall, adding to warming. In addition, low level clouds are thought to have a cooling effect and high level clouds are thought to have a warming effect. The latest satellite data suggests clouds in general have an overall cooling effect. It is the whiteness of clouds that cools us by reflecting energy back into space.

If the overall combined cloud and water vapour effects are not as positive as they are thought to be, this could limit warming to tolerable levels.

A recent Hungarian analysis (44) suggests that the cooling effect of clouds and resulting rainfall will totally negate effects of increased CO2 and keep the greenhouse effect at a constant level. So far there appears to be little general scientific support for this analysis. Arguments put forward to support this theory include suggestion that global relative humidity levels are falling, thus offsetting CO2 increases and thus supporting the theory. However this suggestion doesn't appear to match other current humidity data from different parts of the world, which shows global humidity increasing.

Oceans might absorb the additional CO2

Most of the earth's moving CO2 is held in the oceans. A greater amount of carbon again is held in the earth's mantle and below this.

It is estimated that around 50% of human CO2 emissions are currently being absorbed by the oceans.

The effects of the oceans on atmospheric CO2 levels are also not well understood. There is debate about the ocean's continuing ability to absorb additional gasses. Oceans absorb gas when they are cold, and out-gas when they are warm. Our oceans are warming. Some scientists suggest this absorption of CO2 from the atmosphere is diminishing.

If ocean warming is being caused by something other then CO2, then the warming of the oceans and subsequent out-gassing could be causing the current rise in CO2, rather than humans. There is however no evidence to support this. The previously described carbon isotope analysis suggests that additional CO2 is of fossil fuel origin.

If CO2 or other factors cause warming of the oceans, ocean out-gassing could then also occur, thus amplifying overall warming effects.

There are also a number of other "feedback" effects that could amplify expected levels of warming, such as reduced energy reflection from reducing sea ice, and increased CO2 and methane out-gassing from melting arctic tundra regions.

Maybe it's the sun

Skeptics will often argue that increased warming could be caused by cyclic solar output variations, rather than increased CO2.

Mainstream science agrees that variation in solar output has indeed caused climate change in the past, such as the "Maunder Minimum" of the "Little Ice Age", which was an unusual period of virtually no sun spot activity, and also the temperature rises in the first half of the 20th century.

The sun has some significant cyclic variations, with observations of 11, 22, 87 and 210, 2300 and 6000 year cycles (Wikipedia: Solar Variations), and with a number of these cycles currently reaching peaks. Solar cycles are thought to have driven the cyclic climate changes over the past 1000 years, including the "Medieval Optimum" and minimums of the "Little Ice Age". However, the effects of these are thought to be limited to 0.1 to 0.2 % changes in solar activity and temperature changes of less than 1.0 C.

In contrast to older reconstructions, most recent reconstructions of total solar irradiance indicate only a small increase of only about 0.05 % to 0.1 % between the "Maunder Minimum" and the present. (source Wikipedia: Solar variations).

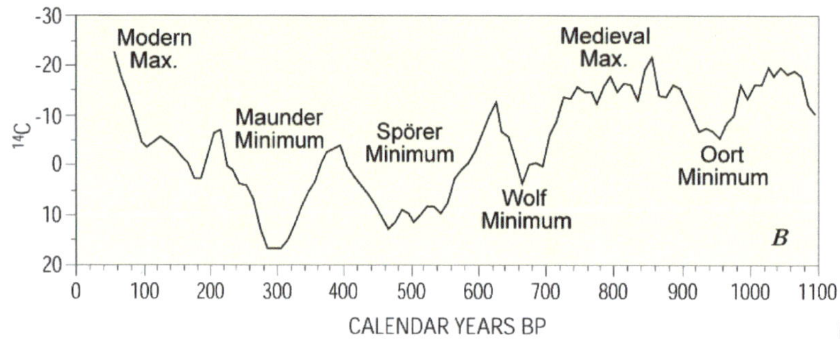

(12)

The above graph shows changes in carbon-14 concentration in the Earth's atmosphere, which serves as a long term "proxy" measurement of solar activity. The amount of carbon 14 that is produced naturally in the upper atmosphere changes slightly with the level of solar activity. The graph matches these variations with known warm and cool times.

There is debate as to whether the sun's output has changed much over the past 30 years. If there is any change it is very small. The charts below of satellite measurement of the sun's normal 11 year sun spot cycles since 1978 show no clear overall rising pattern.

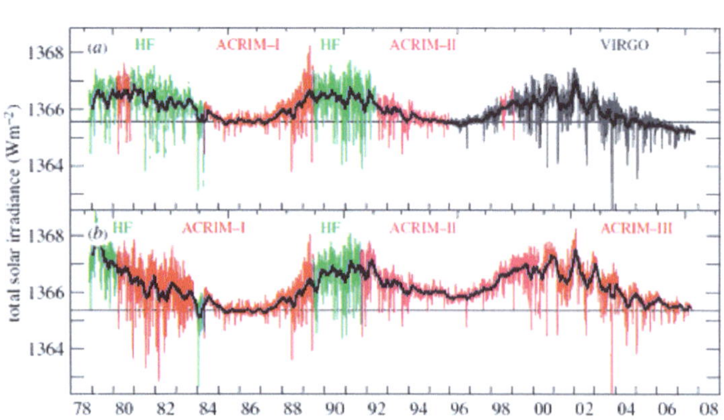

(13)

PMOD TSI composite (top) versus the ACRIM TSI composite (bottom). Coloured lines give the daily values with the black solid lines giving the 81 day mean.

The general scientific consensus is that solar variations have had only a moderate effect over the past century and will also do so through the next century.

The estimated effects of changes in solar output over the last century on temperature are shown in the "Climate Change Attribution" diagram on page 5.

Here are some relevant calculations. It is estimated that the sun's output has increased over the last century by 0.25%. This is part of a number of long term cycles of the sun. This in itself would provide a very significant additional warming force if compared against the sun's total average vertical power of 1366 watts/m2. However this figure reduces significantly for the following reasons. 30% of the sun's energy is reflected back to space. Also, whilst greenhouse gasses have a 24 hour effect across the entire globe, the sun's effect is lowered because it operates on each part of the earth for 12 hours a day. At the poles and early morning the angle of the sun's incidence lowers its power even further. As a result of these effects the sun's average vertical power of 1366 watts/m2 lowers to only 240 watts/m2 on average across the earth. The increase in power of 0.25% across the last century is thus effectively less than 0.5 watt/m2 (0.25% X 240 watts/m2). This is a fair bit smaller than the estimated effects of current CO2 increases so far (1.6 watts/m2).

In summary, skeptics argue, probably correctly, that solar variations have caused the significant changes in climate over the past 1,000 years. It can be seen that historical temperature changes have followed changes in solar activity. This is all probably true. The counter message from mainstream science is that the warming effect of additional CO2 could be much greater than any of these solar effects seen so far.

As stated previously, there is evidence to confirm that recent warming of the earth is being caused by increase in greenhouse gasses. Different causes of warming have different "fingerprints". Reasons to suspect an increased greenhouse effect as the main cause include observed warming at night time, warming at the poles, warming of the troposphere (lower portion of the atmosphere) and cooling of the stratosphere (higher portion of the atmosphere).

CO2 levels were much higher in the past

Skeptics argue the CO2 levels were much higher in the earth's distant past – up to 20 times higher than current levels - even in colder times, so CO2 does not historically appear to drive temperature.

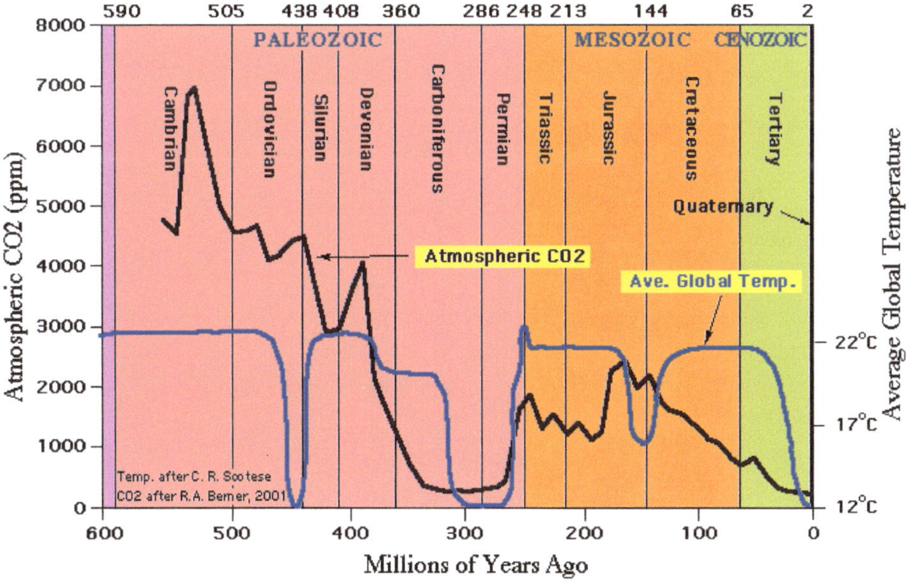

(14)

The mainstream scientific view argues that the world was a different place in the distant past. Solar activity was 5% less than now and the shape of the continents and oceans were much different.

Solar variances and greenhouse effects both impact temperature.

Atmospheric CO2 levels over this very long term have been dropping. Plant and animal life has used it and gradually sequestered it in a fossilized form in the ground.

Countering this, the sun has been getting gradually hotter and bigger. Many millions of years ago it was a fair bit cooler than now, when CO2 levels were higher. In billions of years it will become a red giant and swallow us up. This is a long term effect and very gradual.

The distant past figures of 5,000 ppm CO2 levels or more look very high. Yet this is only four "doublings" of current CO2 levels (600, 1,200, 2,500, 5,000). It is thought that each "doubling" has an equal effect on temperature. We are currently facing a likely prospect of one or two doublings of CO2 by the end of this century, at the current pace of emissions growth. The graph shows that previously it took millions of years for this sort of change to occur.

For the first time ever, we will have the hotter sun and the increased atmospheric CO2.

Historically, increasing CO2 levels followed temperature rise, not caused it

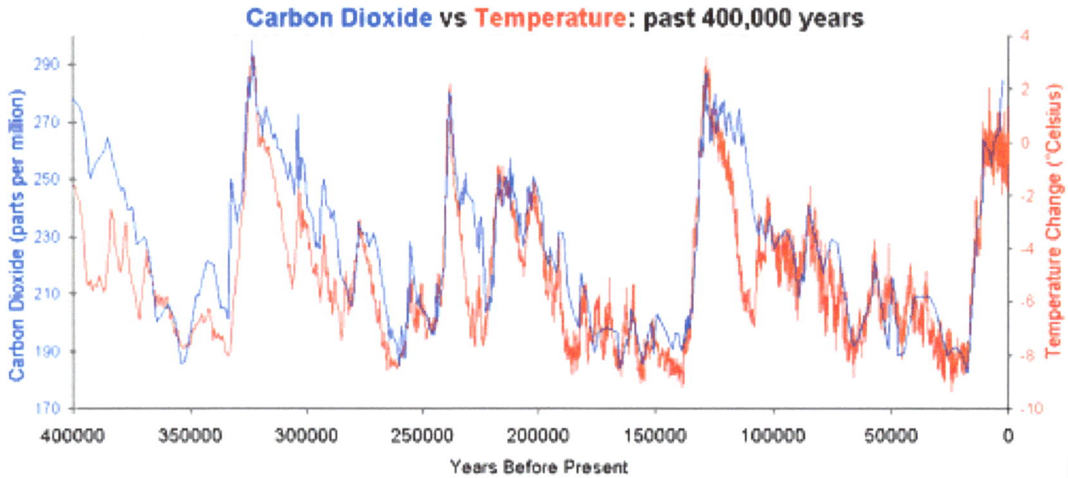

(15)

Ice core data shows that CO2 levels and temperature estimations are correlated over the past several hundred thousand years. However, scientists have observed that the ice core data suggests CO2 level changes actually followed temperature changes by about 400 years.

A reason for this might be that oceans, which store most of the earth's moving carbon, release their dissolved gasses after they warm.

The mainstream scientific view agrees that ice ages probably ended because of gradual changes in solar strength, and then following this atmospheric CO2 levels increased, perhaps from ocean out-gassing, and this helped further to warm the earth.

There is also a counter-suggestion that the accuracy of these ice-core measurements can be misleading, and that some slow exchange of gasses between the ice and the atmosphere continues for a number of years, and this could cause results to look as above.

Mainstream scientists would also argue that even if CO2 levels followed temperature changes in the past, this doesn't mean they couldn't lead them now and in the near future.

Recent temperature rises are not unprecedented

In his 2009 presentation "Apocalypse ? NO!", climate change skeptic Lord Monckton presents a graph of temperature changes since 1659 recorded in central England. This is known as the Central England Temperature Series and is one of the very few long term weather records taken with actual thermometers. To demonstrate that recent temperature rises are not unprecedented, he points out that temperatures rose about 2.2 degrees in 30 years after the "Maunder Minimum" of the recent "Little Ice Age".

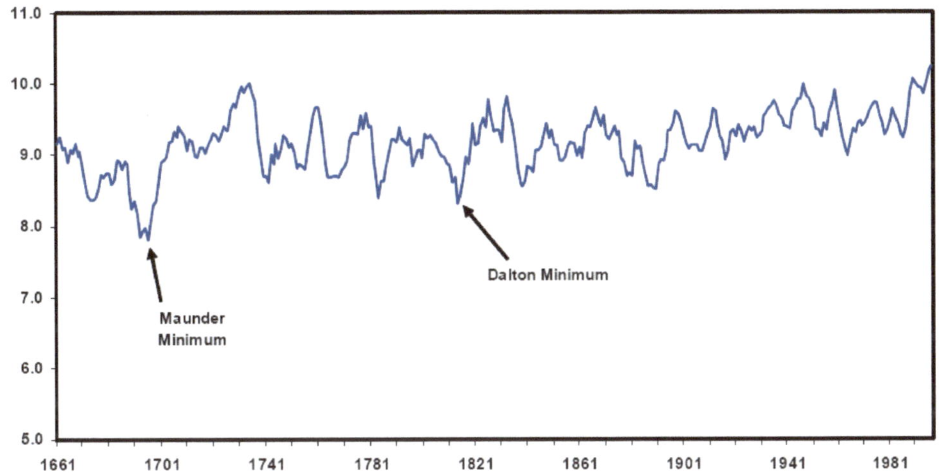

(16)

Central England Temperature Series.

The source references for this data (16) state that for various reasons the temperature records at those times are not totally reliable, and rounded temperatures to the nearest half degree, but still the overall change reported is probably roughly true. It is known that the "Maunder Minimum" during the "Little Ice Age" coincided with a period of very little solar activity. It thus shows that recent temperature rises are not unprecedented, at least at this one part of the world.

Regardless of this, the overall graph is similar to the much criticized IPCC * "hockey stick" graph, that suggests the last 50 years represents not only an unusual period of warming, but also an unprecedented level of warmth.

Other similar graphs that incorporate wider sets of data recorded at a broader set of locations also suggest the last half century is both an unprecedented period of warming and warmth.

This argument over what has occurred in the past is not necessarily of relevance to discussion of whether increasing CO2 will have an increasing warming effect into the future.

* The IPCC is the Intergovernmental Panel on Climate Change, the United Nations organisation set up to oversee climate change science.

Maybe it's a natural long term cycle

In the last glacial period, on a 1,470 year cycle, but not always occurring, there was a naturally occurring sudden global warming that is clearly shown in ice records, but is of unknown origin. These are called **"Dansgaard-Oeschger"** events. A large temperature rise of about 5-8 degrees occurred over just a few decades. It follows and is followed by periods of gradual cooling. The last one was about 11,500 years ago.

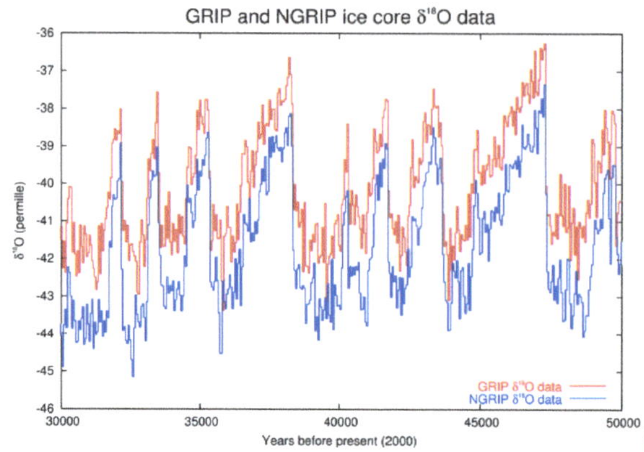

(17)

From Wikipedia. *"The events appear to reflect changes in the North Atlantic ocean circulation, perhaps triggered by an influx of fresh water. The events may be caused by an amplification of solar forcings, or by a cause internal to the earth system - either a "binge-purge" cycle of ice sheets accumulating so much mass they become unstable, as postulated for Heinrich events, or an oscillation in deep ocean currents. The little ice age of ~400 to 200 years ago has been interpreted as the cold part of a D-O cycle, putting us (even without the effects anthropogenic global warming) in a period of warming climate. Dansgaard et al. speculate that these may be related to quasi-stationary modes of the atmosphere-ocean system. D-O Events tend to be what drives the "Sahara pump" which has had an effect upon human evolution and dispersal."*

More recently occurring since the end of the last ice age are similar **"Bond events"**, which have much smaller temperature variations of less than 0.5 of a degree, and are also thought to occur every 1470 years +/- 500 years. These are shown by the black line below.

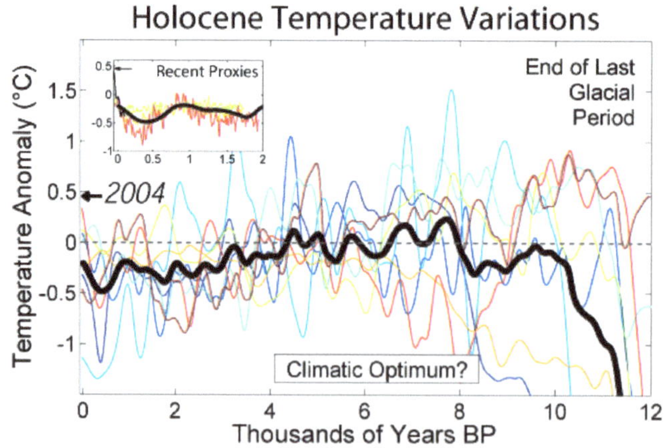

(18)

The last Bond event was 1400 years ago, meaning we are possibly in another one now.

However the existence of these events, does not mean humans are not also warming the planet, in addition to these natural cycles. No-one has thrown all the fossilised carbon fuel sources into the air before. Looking at the temperature scale on the left of this graph, the changes in temperature are very small. The forecast temperature rises over the next century would push the above graph over its top level in a matter of just decades, which would certainly be an unprecedented event, and not one the world could slowly adapt to, like the very small and slow changes shown in this graph.

Maybe it's cosmic rays

There are theories that the Earth's very long term temperature changes could be caused by our solar system moving through spiral arms of the milky way galaxy, with resulting changes in the level of cosmic rays (which, interestingly, are reported to average at 6 particle per cm2 per second).

However the path of the solar system around the galaxy is very slow and thus not likely to cause rapid climate change.

A key argument that global warming over the past decades is not linked to cosmic rays is comes from evidence that the cosmic ray measurements over the past 50 years do not show any trend. A study by Schiermeier, 2007 (19), showed that global warming since 1985 has been caused neither by an increase in solar radiation nor by a decrease in the flux of galactic cosmic rays.

Maybe we've stopped warming

Skeptics will sometime argue whether the Earth is actually warming, which depends on the timeframe considered. For example:

Over the past 10,000 years, since the last ice age, it has been warming

Over the past 1,000 years, it has been mostly cooling

Over the past 100 years it has been mostly warming

Over the past 10 years only, as a short term trend, it was thought it was possibly cooling.

To a mainstream scientist, these arguments are not necessarily relevant to the analysis of what might occur over the next century.

Skeptics have recently argued that the Earth has not warmed over the past 10 years. This view is not well supported by the following graph, which includes most recent data, and is still consistent with an overall rising trend.

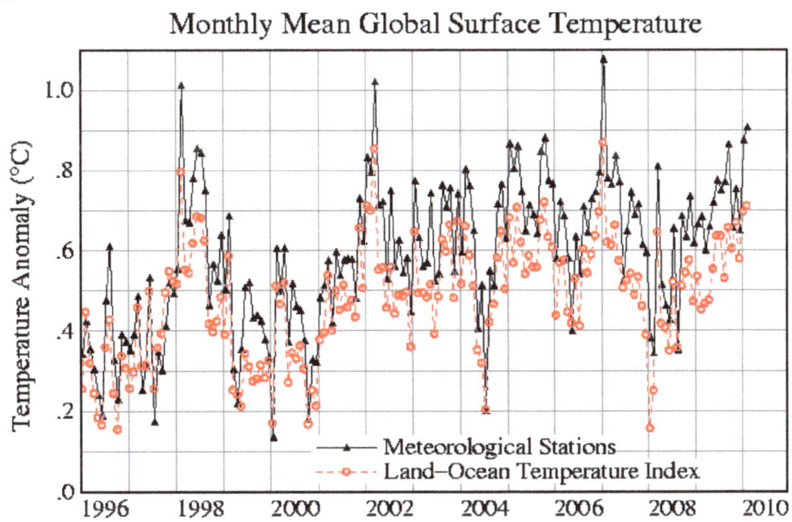

Source http://data.giss.nasa.gov/gistemp/graphs/Fig.C.lrg.gif (11)

In his 2009 presentation "Apocalypse ? NO!" climate change skeptic Lord Monckton criticizes mainstream science for presenting graphical evidence of warming which distorts the view by assuming a certain start point. He then presents the first graph below, regarding sea temperatures at the Australian Great Barrier Reef, which has suffered from coral bleaching particularly since 1998. Next to it is shown a different and longer term view.

"Barrier Reef is just fine"

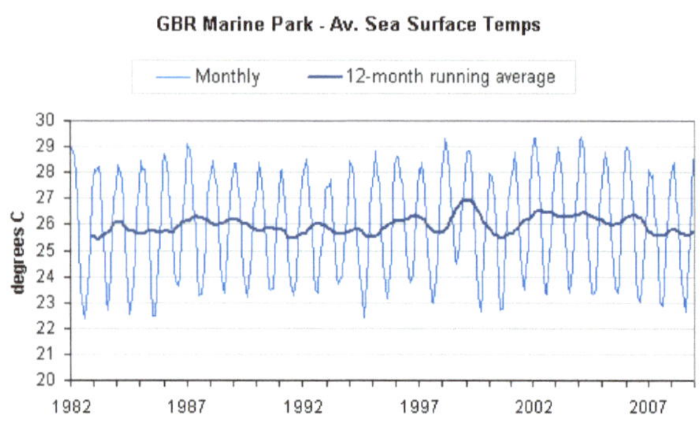

(27)

A different, longer term view

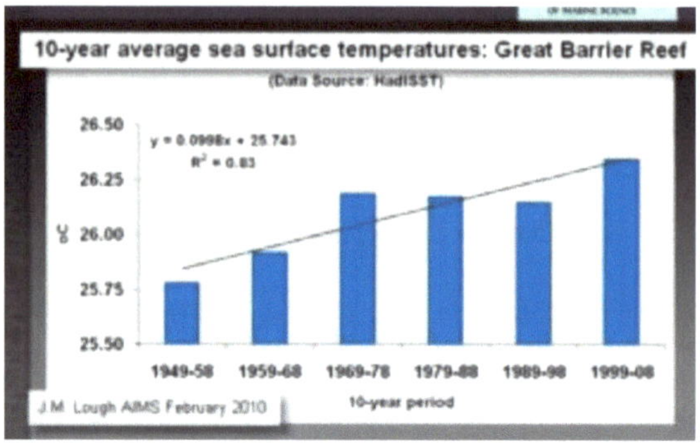

(28)

Mainstream scientists say that short term 5 to 10 year cyclic effects should not be considered in considering overall patterns, and that a view of at least 25 years is required.

The most likely truth

Skeptics put forward many counter arguments, but this closer examination suggests these arguments do not hold much water. None of the alternative theories as to the likely cause of recent warming appear to stack up against the mainstream scientific conclusions that recent global warming is caused by human activity that is adding CO2 to the atmosphere.

In the end, one need only ask the following four questions:

Is CO2 a significant greenhouse gas?

Are CO2 levels in the atmosphere increasing?

Are humans probably causing this?

Could an increased greenhouse effect warm the planet?

The intuitive answer to all of these questions, based on current evidence, is unfortunately, yes.

What one can reasonably hang on to, as a skeptic, is a hope that the effect on temperature of increasing CO2 levels will be small. Again, intuitively this does not seem likely, given the effects observed so far with the rise in CO2 of only around 30%, and in the context of now rapidly rising CO2 emissions around the world, especially from developing countries in the context of an increasingly global economy.

Larger effects on our climate are thus likely in coming decades. It is estimated that we will see a 3 degree temperature rise at double pre-industrial CO2 levels, which are likely to be reached before the middle of this century. This will likely result in about 4 times the temperature rise seen so far, of only around 0.7 degree. Even greater climate effects again are likely to be seen if CO2 levels continue to rise beyond that, and possibly even if they could be stabilised at such a level.

The risks of warming

Effects of warming, with a modest 0.7 degree rise, can be seen now. These include record temperatures, reductions in arctic ice, changes in rainfall, melting of glaciers, slight increases in sea levels and slight reductions in snow levels.

The risks of further warming to the world in general are discussed in IPCC documents. The risks to Australia are discussed in the Garnaut review (66) and by Pearman (39). At higher levels of temperature rise the effects are forecast to be severe.

A detailed discussion of the forecast effects of higher levels of warming in Australia and the world as a whole are well documented but are not replicated in this book. This is done in order to keep this book easy to read.

The forecast potential consequences of climate change far exceed the costs of fighting climate change. In some circumstances, they cannot be compared at all.

The Australian Garnaut climate change review undertook an extensive analysis of climate change risks and potential costs for Australia. It suggests costs of climate change can be analysed in four categories.

1. Economic costs that are readily estimated, such as impact on agriculture
2. Economic costs that are more difficult to forecast and estimate, such as impact on tourism, or from increased international conflict and/or refugees
3. Insurance costs against the risk of catastrophic climate change
4. Losses of utility or amenity, such as loss of natural assets that Australian's value, such as other species, the Great Barrier Reef, or Kakadu.

The financial analysis suggests that climate change mitigations costs are far less than the potential economic costs to Australia of climate change towards the end of this century and into future ones.

The Garnaut report says that we will experience economic loss from climate change towards the end of this century and that this cannot now be avoided. The loss is likely to exceed the costs of mitigation. All we can do is try to minimise the loss.

The chances of catastrophic, or "run-away", climate change (type 3) are difficult to quantify. This could occur at certain human-caused temperature levels that start to decay the world's rain forests, or cause the release of CO2 and methane from soils and oceans at levels potentially much higher than that caused by human activity.

In the book "Climate Change Science and Policy" (65), an interesting point is made. We have some level of ability to control human caused CO2 emissions. If temperatures rise to the point where additional natural emissions start to become major, we will not be able to control them. Possibly even the dangerous geo-engineering options (also not discussed in this book) might not work.

In the author's view, it is not just the significant economic benefits and risks that should drive our climate change mitigation thinking. It is because of the risk type 3 that we should be doing as much as possible.

The book ZeroCarbonBritain (67), which advocates more aggressive targets for Britain than this book does for Australia, draws an analogy. It describes the earth's potential feedback effects as "the bomb", and describes human-caused CO2 emissions as "lighting the fuse". We need to try to stomp-out the fuse.

PART B – THE POLICIES

In general

This part of this book looks at current climate change mitigation policy proposals of Australia and other nations. It suggests improvements to those policies and also suggests a better policy that would be more appropriate to address the significant risks of climate change. This is the more important part of this book, because regardless of who is correct in the scientific debate, this is about what we can do to address the significant risks of climate change.

Given the size of the risks, it is prudent that we act aggressively now.

It is forecast that the world will need additional energy in the future. Much of this additional requirement will be in developing countries, as they become more affluent, and resulting from increasing levels of production of cheaper goods for the global economy. Some of this will result from increased air conditioning as heat levels rise. It would also be necessary to have more energy sources if, instead of global warming, Europe were to swiftly enter another ice age, which is thought by some to be a potential result of melting of Greenland ice packs and its effect on currents in the Atlantic Ocean and particularly the gulf stream (35).

We could easily downgrade the urgency of our action if in 10 years time the climate change skeptics are proven right, and we find we do not need to be alarmed. Unfortunately we will probably do this the other way around, where we will wait 10 years and then possibly increase the urgency. Assuming the science is correct, the approach required then will then be more expensive in every way, in that it will become a lot more expensive and also much harder to react in time, and our economic ability to do so will also be reduced due to increasing climatic impacts.

Humans are generally complacent about climate change. We feel we can tolerate temperature rises of just a few degrees. We can just go inside and turn up our air conditioners, and that is probably true. What most people don't yet understand is the effect such temperature rises are likely to have on our environment, other species and issues of great importance to us such as global and local food supply.

Most forecasts suggest we have less than 10 years to react.

What is the world doing?

The Copenhagen Accord was an agreement that the world should aim to limit warming to less than 2 degrees, which is thought to be a level at which we would avoid "dangerous" effects of climate change. In order for this to be achieved, the IPCC has suggested total world emissions need to peak by 2015 and then start reducing.

The Copenhagen Accord is not a legally binding agreement, but most nations have already agreed to take some action and have set emission reduction targets for 2020.

No nations have yet set targets or put policies in place that appear sufficiently aggressive to achieve the overall longer term target, but nevertheless some initial commitments are being made:

Australia: 5% reductions and up to 25% if there is stronger international commitment;

Europe: 20% from 1990 levels and 30% if there is international agreement;

USA: 17% from 2005 levels;

China: a 45% reduction in carbon "intensity";

India: 25% reduction by 2030;

Japan: 25% from 1990 levels;

Russia: 25% from 1990 levels and

Brazil: up to 38.9% reduction by 2020.

The Copenhagen discussion was largely about achieving emissions equity between developing and developed nations. Since then, some nations have put emission reduction proposals on the table and are saying they will increase them if other nations do too. Unfortunately, this "*we will if you will*" approach might not be sufficient to save us, but could be a way of encouraging greater international effort.

To date, Australia's commitment is a target of 5% below 2000 levels by 2020 and up to 25% if there is international agreement to take more aggressive action. Analysis of the Australian Treasury documents suggests that a stronger level of commitment from Australia and the world in general is required to limit warming to 2 degrees and avoid "dangerous" climate change. Australia's lower 5% reduction commitment is thought to correspond with a probable 3 degrees of warming. In practice temperatures and their effects might run higher than currently predicted.

The USA, China, India, Russia and Japan are the world's five largest emitting nations. These five countries represent over 50% of the world's CO2 emissions. All other countries represent less than 2% each. Australia is the world's 14th largest emitter. China and India together are currently nearly half of the world's population, and with their economic growth these two countries are forecast to become the world's largest carbon emitters.

Those five countries are the ones that particularly need to lead the world in climate change mitigation. The author suggests the next attempt at world agreement on unified action should be lead by those five countries, plus Australia.

Can we make a difference?

This is an important question. Skeptics will often argue that humans are having minimal effect on temperature but also as a result, the effect of reduced emissions will be small and not worth major investment. This argument would potentially be valid if very low climate sensitivity is assumed.

In contrast, arguments from the climate change modelers suggests that reduced emissions targets could help cap warming to certain levels. For example, if aggressive policies are put in place the warming could be limited to 2 degrees, which is still about four times the temperature rise over the last century, but is a level at which it is thought we can avoid "dangerous" levels of climate change.

To do so requires actual emissions to start reducing this decade.

Given the balance of scientific argument shown in Part A, the enormous risks, and where the best expertise seems to lie, it seems prudent that we should follow advice from the modelers rather than the skeptics. Given the level of the risks, the cost of implementing an expensive climate change management policy will be much less expensive than economic effects of not doing so. The IPCC put a price on human lives in one of its risk assessments (differing in different parts of the world), and was criticised for doing so.

If we cannot sufficiently limit emissions, and the effects become adverse, we may still be able to resort to climate "geo-engineering" options, where we release other gasses or employ other measures to try to counter act CO2 levels and cool the earth. However, these approaches will also have risks.

The potential for renewable energy in Australia

Energy generation is the activity with the largest potential for achieving emission reductions, now at 50% of our overall emissions. In an Australian Government report on the potential for renewable energy (30), it is suggested a "clean energy mix" in Australia could be delivered by 2040, consisting of:

- coal generation 9%
- natural gas 30%
- biomass using crop residues 28%
- wind 20%
- hydro 7%
- solar only 5%, but contributing significantly during peak times

A more pessimistic scenario discussed in the same report suggests by 2040 Australia will achieve renewal energy sources at only 47% with coal at 30% and gas at 20%.

As part of this new infrastructure development, there will need to be technology advances and investment in energy storage methods. Hot water based systems and other fluids such as "molten salt" are currently being used. Excess energy during the day is used to heat such fluids, or to create hydrogen to provide for a continuous energy supply. In order for solar facilities to meet peak load demands, some facilities begin storing energy when the sun comes up but do not start generating electricity until it is required for peak loads in the latter half of the day and into the early evening.

Geothermal approaches to energy generation include drilling down to "hot fractured rock" at temperatures of 250 degrees at depths of 3-5 Km in order to generate steam. Research suggests this approach could provide a potential energy source to serve Australia's total needs for 20,000 years.

A geographic spread of renewable energy generation facilities assists in creating a national power system, and avoids some of the issue of transmission line losses, which can be significant.

Alongside these new methods of producing energy, new methods of powering transport are required as this contributes significantly to carbon emissions. This can include renewable electricity and hydrogen.

The report suggests that cost of renewal energy is dropping and is expected to become cheaper than the developing "clean coal" approaches which are still under development, possibly within a decade.

The report concludes that:

> "**there are no technical or financial impediments to renewables providing about 50% of all Australian electricity demand by 2040**"

and also

> "**in the longer term, it appears there is no technical reason to stop renewable energy from supplying 100% of grid energy**".

Later chapters of this book explore the costs of achieving 100% renewable energy.

Comparison of Australia's proposed policy approaches

Australia's proposed policy approaches

Australia, like many nations, has a range of policies and incentives to promote renewable energy that have been in place for a number of years. For example, in place is a Mandatory Renewable Energy Target (MRET) for achieving renewable energy generation at 20% of total energy usage by 2020.

For future strategies to address the risks of climate change, Australia is currently being offered two very different options by the main political parties. Both claim to provide for similar emission reductions of 5% by 2020, but the approaches and strategies are so different there are large implications beyond that time frame.

The current Government's proposal is an emissions trading approach known as the Carbon Pollution Reduction Scheme (CPRS), which is an approach that forces industry to act. Due to resistance in Australia to its implementation, and uncertainty regarding the size and nature of the world's response, the Australian Government recently announced a delay in its proposed implementation until 2013.

In the interim, the Government also has in place a number of direct action policies. These include the Mandatory Renewable Energy Target and the Green Energy Initiative.

In stark contrast, the Australian Opposition party's climate change policy is a proposal for direct action by Government.

The Opposition party sells its primarily direct action based approach as a lower cost alternative. The Opposition calls the Government's proposal the "***big new tax on everything***". It claims that its policy will achieve a similar result to the Government's proposed approach for only around 1/10 of the cost.

Despite its proposed delay, the Government's proposed emissions trading approach is the more aggressive and scalable of the two policies, and so is the better one for addressing climate change. However, that doesn't mean it is necessarily the best approach that is available to us. Suggestions for improvements to this policy and also the Opposition's proposed policy are made in this book. A different and more aggressive policy is also proposed.

The various alternative policies that are in place, or being suggested for the future, are described in the following chapters.

Australia's existing policies – The Mandatory Renewable Energy Target, Clean Energy Initiative and Solar Flagships Programme

Australia's **Mandatory Renewable Energy Target** has been extended to achieve 20 percent renewable energy generation by 2020. This will require annual production of 45,000 GWh of electricity, which is 20% of Australia's expected total energy usage at that date.

The **Clean Energy Initiative** provides funding to assist the development of new cleaner energy approaches, of $5.1B over four years. This total funding includes $2B for "clean" coal, $1.6B for the Solar Flagships Programme, and $465M to promote renewable energy research.

The **Solar Flagships Programme** aims to create 1,000 MW of large scale solar electricity generation capacity. Tenders have been submitted for the first stage including both photovoltaic (solar cell) and solar thermal approaches. This is claimed by the Government to be currently the largest solar energy project in the world. (Similar large work is planned in the USA, particularly California, and larger construction contracts have been recently signed in China).

The Solar Flagships Programme will only play a small part in achieving Australia's renewable energy emissions target of 20% for 2020. Like much of the rest of the world, Australia is currently at 7% renewable energy, so we have about 13% to go to meet the 20% target.

Under the 2010 budget $652 milllion was also allocated to support renewable energy projects like wind and solar power as well as to encourage households to reduce their energy usage.

Under the existing Government's policy, this and a variety of other existing programmes of subsidising direct action are interim measures until an emissions trading based economy is potentially put in place and becomes effective.

At this point the reader will have noted that the larger expenditure in the Clean Energy Initiative goes towards "clean" coal, rather than renewable energy. The author suspects this emphasis is tied to Australia's interests in coal mining and export. Australia is the world's largest coal exporter, and coal is Australia's largest export. The "clean coal" approach is discussed further in later chapters. Although development of a truly clean approach for coal-fired energy production would be in Australia's and the world's interests, the most likely methods, which involve pumping CO2 into the ground, are not the best and cleanest way for us to go. This is discussed further in later chapters.

The Government's proposed Emissions Trading Scheme

The Australian Government's proposed **Carbon Pollution Reduction Scheme** (CPRS) is an emissions trading scheme where companies pay for "permits" for greenhouse gas emissions. These can, if necessary, be topped up by purchasing permits from other companies that no longer require them, or otherwise by purchasing "offsets". Alternatively, instead of purchasing emission permits or offsets, companies can elect to move to lower carbon production methods if that is cheaper. The cost of carbon emissions will gradually force industry to find lower carbon approaches and to restructure.

Carbon trading schemes are already in operation in Europe and in Japan and are being considered by other countries such as the USA.

In order to bring down emissions the total national emission allocation will be capped. Carbon permits will be auctioned. This is called a "cap and trade" system. The cap on emissions will reduce over time in line with the planned emission reduction targets. As the number of permits reduces, the market price of emissions will gradually increase and emissions will gradually come down.

Carbon "permits" and "offsets" are similar things. Both can be bought or traded. The former is a permission to emit a tonne of carbon. The other is an offset purchased from another organisation that is undertaking an emissions reduction activity that can be purchased to counteract any additional emissions.

If additional permits or offsets are required, these may be purchased on the international market, meaning that some of the emission reductions will actually occur overseas instead of in Australia. Overseas carbon permits or offsets will be purchased if they are cheaper, resulting in lower costs to industry. A cap on international trading will be set so that at least 50% level of investment occurs in Australia. These emission reductions purchased internationally are still considered part of Australia's emission reductions.

A level of "shielding" and subsidies will be provided to protect large emitters in early years, as well as to households to subsidise a small rise in annual costs. These subsidies in initial years will be provided from the funds gained from the auctioning of permits.

The effect on Australia's total gross domestic product is estimated to be small, with various estimates from the Australian Treasury Department (43) around 1.5% p.a. Costs estimates per household are also relatively small, of the order of a few hundred dollars per year. These estimated costs are both within Australia's future expected GDP growth and growth in "real" household income respectively.

The scope of the scheme covers around 75% of Australia emissions. It is proposed that agriculture be excluded from the scope of the scheme, to protect that sector. Agriculture is approximately 16% of Australia emissions. There is debate around how long this exclusion should last.

Revenue from auctioning permits in early years is estimated at $13B, which seems to the author to correspond roughly to Australia's total current 550 Mt of emissions at a price of $23/tonne (the author has requested a description of how this figure is derived from the Department of Climate Change). This annual revenue is expected to grow over time.

Also to make the initial year easier, the Australian government is proposing that trading in the first year will be fixed at only $10 per tonne, with no caps, and overall revenue around $4.5B (42). Once real trading commences, a cap of $40 per tonne will be set, and this will rise with CPI.

The Government's policy clearly forces industry to find the best low carbon approaches and will increase operating costs. Critics of the policy say it will make Australian industry less competitive, cause large job losses, and force investment overseas, all due to higher operating costs in Australia. The approach could result in transfer of some business operations and carbon production overseas ("carbon leakage"). Also, where permits or offsets are purchased overseas, this also pushes some investment in new green industries overseas.

A variety of different emission reduction trajectories have been examined. These different scenarios have different levels of emissions caps and therefore carbon pricing, from the least aggressive reductions and lowest carbon price (CPRS-5), to most aggressive reductions with the highest initial carbon price (Garnaut –25). The former matches the proposed current emission reduction target of 5% by 2020. The latter is the higher level of commitment required to meet Copenhagen target of limiting warming to 2 degrees, but achieving this also assumes an increased international effort.

The following table summarises the alternative carbon trading scenarios, which provide the different emission reduction trajectories, as outlined in the Australian Treasury papers.

Scenario name *	No climate change mitigation	**CPRS – 5**	CPRS – 15	Garnaut – 10	**Garnaut - 25**
Starting Carbon Price/tonne	0	**20**	28	24	**43**
Emissions reduction in 2020		**5%**	15%	10%	**25%**
Forecast Results in 2050					
CO2 ppm	1560	**550**	510	550	**450**
Temperature increase limit (*50% chance – Author's addition*)	8+	**3**	2.6	3	**2**
Global emissions reduction per capita	0	**-38**	-44	-41	**-66**
Australian emissions reduction per cent	0	**-60**	-60	-80	**-90**

* The different scenarios are named after the level of emission reductions achieved in 2020. e.g. CPRS 5 achieves 5% in 2020. Two of the scenarios are marked in bold. One is the CPRS-5 scenario which matches Australia's current proposed 2020 emissions reduction commitment of 5%. The other is the Garnaut-25 scenario, which matches our potential higher 2020 commitment of 25%.

The Australian Treasury papers (38) explained, at the time they were written in 2008, how the planned emissions trajectory would be chosen as follows: "*The emission caps for this scheme will be set in 2010, in light of the national emissions trajectory and targets and final decisions regarding scheme design. The Government will take a range of factors into account in setting the trajectory and targets, including the work of the Garnaut Climate Change Review, modelling in this report, consultation with stakeholders and the international negotiations now underway.*"

The following graph shows the planned emissions reductions of these different scenarios through to 2050. It shows that CPRS 5 and 15 achieve a similar end result, but CPRS 15 starts lightly more aggressively. The upper black line shows Australia's growth in emissions if no further action is taken.

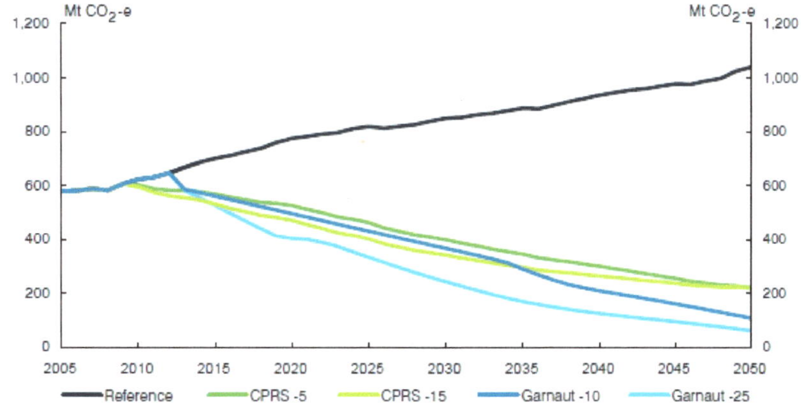

(38 – Reproduced with permission from the Commonwealth of Australia)

The Australian Treasury papers (38) also make the following comment regarding CO2 stabilisation at 550 ppm, which corresponds to the modest 5% emissions reductions currently proposed under the CPRS – 5 scenario.

> ***"Between 20 and 30 per cent of all species are projected to face a 50 per cent likelihood of extinction under this scenario (IPCC, 2007b). This would involve the total realignment of ecosystems across Australia. Coastal communities, agriculture and infrastructure would all face significant risks of adverse impacts. These include frequent or permanent coastal inundation for parts of the Australian coastline, a substantial increase in extreme weather across the nation, and substantial restructuring of the rural sector (Pearman, 2008)."***

This would suggest that our emission reduction targets really need to be greater than the CPRS-5 scenario, but that is the one currently on the table for Australia while Australia waits for greater international commitments. Could an Australian Government, or the Opposition for that matter, really only commit us to such a low quality outcome? The source report referenced above (39) also describes some possible worse climate change impacts than those summarized above.

Unfortunately the low targets we are currently setting, along with the rest of the world, could thus lead us to huge economic and health impacts, which could actually reduce our economic ability to put in place the measures that are really required. The Australian Treasury economic analysis of emissions trading scenarios states that their economic analysis does not take into account the negative economic effects of climate change itself.

The next diagram from the Australian Treasury papers shows more detail of the anticipated CPRS-5 scenario emission reductions.

CPRS -5 scenario

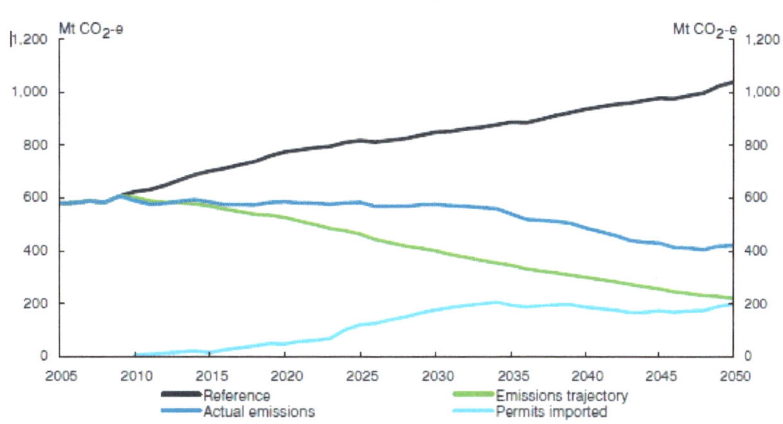

(38 – Reproduced with permission from the Commonwealth of Australia)

It is interesting to note that in this CPRS-5 scenario actual emissions in Australia do not reduce until after 2030, and remain higher than the overall emissions trajectory (green line - actual emission reductions overall), due to the anticipated amount of emissions reductions that are purchased overseas (imported permits - up to 50% allowed).

An overall problem with selling emission trading to voters and industry is, as well as the large scale of the proposal, that the initial and future carbon price are not known, and this obviously creates uncertainty for industry and households as to what the actual cost will be. One reason the price is uncertain is that it depends on international and Australian commitments that will be made as a result of ongoing post-Copenhagen negotiations. The Australian Government has offered a 25% reduction target for 2020 (i.e. the Garnaut-25 scenario) if other nations commit to more aggressive action as well. This uncertainty regarding costs is likely to push many voters away from the Government and this proposal.

The following diagram shows where emission reductions are expected to occur. It is important for the reader to have a good mental image of this picture in order to understand alternative policy proposals that are discussed later in this book.

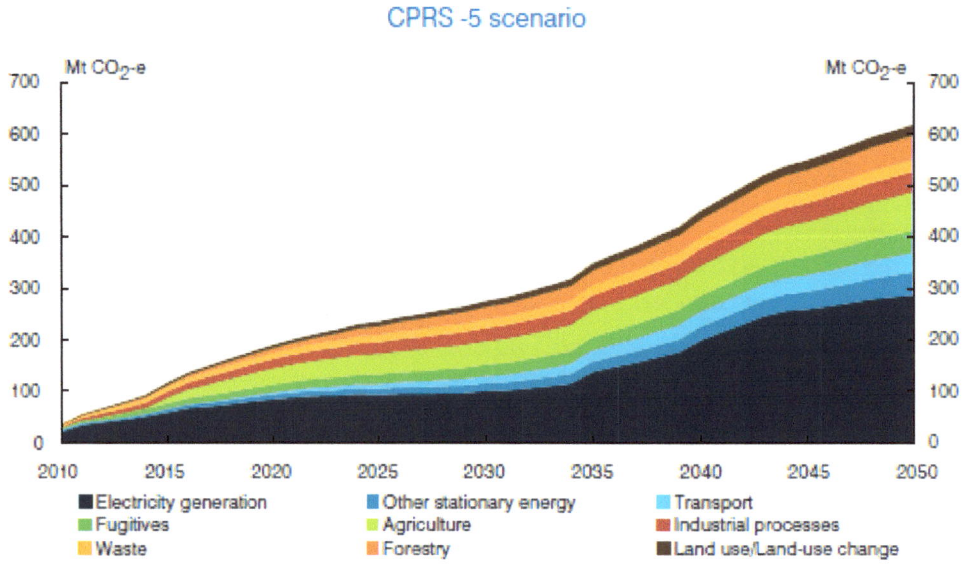

(38 – Reproduced with permission from the Commonwealth of Australia)

The Opposition's Proposed Direct Action Policy

Australia's Opposition party calls the Government's proposed emissions trading scheme "***a big new tax on everything***" and is proposing a much lower cost programme of direct action (41), with a cost portrayed at only about one tenth of the Government's proposed programme. The overall proposed programme is as follows.

	2011-12 ($M)	2012-13 ($M)	2013-14 ($M)	2014-15 ($M)	Total ($M)
Emissions Reduction Fund	**300**	**500**	**750**	**1000**	**2550**
One million solar roofs	100	100	100	100	400
Clean energy Employment hubs	15	15	15	15	60
Solar towns and solar schools	25	25	25	25	100
Geothermal and tidal towns	10	15	15	10	50
Renewable fuels	5				5
Greenhouse friendly programme	2	2	2	2	8
Urban forests and green corridors	5	10	15	20	50
Total	**462**	**667**	**992**	**1172**	**3223**

The **Emissions Reduction Fund** is the main component of the policy and will provide an average investment of $1.2B per year through to 2020. As shown above, it starts at $300M in year 1, reaching $1B in year 4 and would gradually increase to higher levels to achieve this 10 year average.

As well as the fund, the overall policy also lists other direct action initiatives in the first four years, including $100M p.a. for one million solar roofs, and other smaller scale initiatives growing collectively to $70M in 2014-15 for action such as solar towns and schools, renewable fuels, urban forests and green corridors. The Emissions Reduction Fund is however the main proposed expenditure.

The Opposition promotes their policy as low cost, at only $3.2B in total over first four years. Annual expenditure figures after 4 years are not clear, however an approximation of the Emissions Reduction Fund expenditure that would provides the stated average of $1.2B through to 2020 is constructed in the table below. The first four numbers are from the policy document and the following numbers in italics are example estimates that would achieve a $1.2B average.

2011	2012	2013	2014	2015	2016	2017	2018	2019	2020
$0.3B	$0.5B	$0.75B	$1B	*$1.2B*	*$1.3B*	*$1.4B*	*$1.6B*	*$1.9B*	*$2.1B*

By comparison, the CPRS is reported by the Opposition as much more costly, at $40.6B over first four years. The reason these costs look so different is discussed in a following chapter. This is not comparing "apples with apples".

The Opposition's direct action policy is targeted at providing a low cost alternative to the Government's emissions trading minimum scenario (CPRS-5), which targets a 5% reduction in emissions by 2020. Planned reductions at 2020 are therefore 140 Mt (Megatonnes) of CO2, to reduce anticipated emission levels of 664 Mt p.a. in 2020 to around 520 Mt, which is 5% below Australia's Kyoto baseline 1990 level of 547Mt p.a. The Opposition party uses 1990 levels as its 5% reduction baseline, whereas the Government uses 2000 levels. They are not much different.

The following example of how the Emission Reduction Funds might be used to achieve the planned 140 Mt reduction is provided in the Opposition's policy document.

	Potential Available Additional Annual CO2 Reduction by 2020		**Indicative CO2 Reduction to be delivered through Funds in 2020**	**Indicative CO2 price per Tonne**
	Min	Max		
Soil Carbons	150mt	355mt+	85 mt	$8-10
Electricity generators and industry	10mt	30mt	55 mt	$25-35
Forestry Measures	15mt			$15
Waste coal mine gas	4mt	8mt		$10-20
Transport	3mt			$40
Green buildings/energy efficiency	20mt	30mt		$10-15
Landfills/Compost/recycling	6mt	9mt		$10-15
Total	208mt	450mt	140mt	

The suggested main avenue for achieving the total 140 Mt of emissions reduction in 2020 is via an 85 Mt reduction from "soil abatement", where improved farming practices are used to increase carbon storage in soil. At $10 per tonne this would cost approximately $850M.

This approach is expected to be complemented by a range of other measures involving electricity generation, forestry, transport, energy efficiency and waste management. These further reductions will be purchased at a variety of expected higher costs in the $10 to $40 per tonne range to give a further 55 MT reduction. At an average rate of $20 per tonne, that would be a further $1.1BM of expenditure. This results in a total cost in 2020 of around $2B, approximately matching the estimated size of the fund in 2020 that was calculated previously.

The proposed approach for selection of actual emission reduction approaches and management of the Emissions Reduction Fund is that organisations will bid for funding, for otherwise unfunded carbon reduction measures, and projects will be selected on competitive merit.

The Opposition's policy is clearly a softer approach for industry in general. It puts no additional cost on existing operations and provides little encouragement for restructuring. As part of their much smaller scale direct action plan, all investment is kept in Australia.

Unlike the Government's overall policy proposal and position, no suggestion appears to be made in the Opposition policy documents that Australia's emissions reduction commitments, post the Copenhagen discussions, might increase to more aggressive targets than a 5% reduction by 2020.

The soil abatement approach proposed by the Opposition might indeed be a cheaper alternative in the short term, and is a way of extracting carbon from the air to reduce net emissions instead of reducing actual emissions. This is a fundamental difference in policy and strategy between the two parties. The Opposition policy says that it has been estimated, by interest groups, that with improved farming practices Australia could pull up to 355 MT or more of CO2 out of the air annually, on an ongoing basis, with the added benefit of improved soil fertility.
If we can do this, we should. There are also other methods of carbon sequestration being examined around the world. There is however skepticism whether the soil based sequestation approaches can work and whether their results can be accurately measured.

In this author's view, it does not seem reasonable that we can keep emitting more and more carbon each year and expect that we can carry out other ongoing activities each year to pull it out of the air. If Australia or the world does not take effective action to actually reduce emissions, the amount of carbon that farmers will need to extract out of the air annually just grows and grows. The land area and resources to do so would eventually run out, or we might find that the process does not work at all.

The author would assume that the Opposition also does not necessarily see this approach as a long term solution to meeting total carbon reduction targets. This is not easy to assess, as the Opposition's policy does not discuss longer term targets.

Nevertheless, a soil abatement based approach is discussed at some length in Australia's Garnaut report on climate change (66) and it does appear to have strong potential as part of a better constructed long term policy, such as the one put forward later in this book.

Outside of the Emissions Reduction Fund, the next largest suggested policy funding is for subsidising "one million solar roofs". This might sound good. However, if these facilities averaged 2KW each, the one million roofs would generate only 2 GW of power or about 3.5 TWh of energy pa (based on 1500 hours of sunshine per annum). This is only a bit over 1% of Australia's total 257 TWh electricity usage per annum. It could be argued that this is barely worth doing, and in terms of its actual total cost to the community, is not an efficient usage of funds. Before subsidies, these installations currently cost around $8,000 (or more), and so at least $8B would be spent by the community in total to achieve the 2 GW of power. This is an inefficient way to create renewable energy, as we will also discuss later in this book.

The Opposition's policy thus promotes itself as a cheap approach to reduce net emissions for the short term, by 2020. It could be assumed that other different longer term approaches would need to be identified and employed later to reduce actual emissions. This is not however how the policy is being sold and communicated to the public. It is mainly sold as a way of avoiding a "*big new tax on everything*".

The Opposition's policy does not describe any real long term climate change mitigation strategy. It is could therefore be seen as just a "wait and see" or "Claytons" climate change policy. (As an explanation for some readers, Claytons is a non-alcoholic drink that was once advertised as "*the drink you are having when you are not having a drink*".)

The Opposition's proposed direct-action policy does have some potential, if it were scaled to an appropriate size, and also presented a long term plan. A larger scale direct-action plan is presented later in this book.

Is emissions trading efficient and effective?

Some economists describe emissions trading as the most efficient and effective approach to climate change mitigation. Its key principle is that it allows industry to seek out the most cost effective approaches to emission reductions available at the time (a bit like the Opposition Party's proposal). The largest emissions trading scheme in operation is the European scheme. Some analysts declare the European scheme a success, and others see it as a massive failure.

As a result of negative aspects of the European system, some left wing groups, environmentalists and economists voice strong opposition to emissions trading. There is suggestion in other documents (45,46) that emissions trading programmes, such as the European programme, that have been put in place since Kyoto agreement have done little to reduce actual emissions. It is suggested that emissions trading actually holds back climate change mitigation.

The following types of mismanagement are described in these discussions of the European programme.

- There are reports that a common result of international emissions trading is that big emitters in affluent countries continue to maintain their emissions by simply paying for cheap carbon offset projects in poorer countries, such as renewable energy projects in villages, and that the net result is no reduction in emissions at all.
- There are reports that some or many emissions trading activities lead to an actual increase in emissions. A cited example is a company that manufactures refrigerants, spends a few million dollars to reduce the percentage of emissions, then makes hundreds of millions of dollars from carbon credits. It then emits more and more greenhouse gases because of the profits from carbon credits, rather than any need to carry out the original business itself.
- There are reports of villagers being forced off their land for the purposes of carbon emission projects in poor countries, where the project itself has little real value.
- It is suggested that in the European scheme most proposals for carbon offset projects are approved. One author suggests that *'If you are a good storyteller you get your project approved. If you are not a good storyteller you don't get your project through*".
- There are reports of emissions trading funding projects to change business activities or make technology upgrades that would have been made in any case.
- It is reported that as a result of subsidies and over-allocations of permits across different nations, the price of carbon per tonne recently reduced to only EU$1 (46).
- There is suggestion that the emissions reductions reported from emissions trading are often only reductions on paper, and not reductions in reality.
- It is suggested that because Kyoto carbon accounting standards do not recognise the benefits of stopping deforestation, and also to some extent reforestation, this is preventing effective action in these substantial areas of potential world emission reductions.
- There is suggestion that renewable energy technologies such as solar power are not getting much investment from emissions trading, because they are not the currently cheapest option. It is suggested that with greater levels of investment, they would be. There is argument that emissions trading actually harms potentially more effective climate change mitigation methods and efforts in a manner such as this.

This last point above is very significant. All of these points illustrate how emissions trading is not necessarily at all efficient, and that market forces will tend to take cheap, expedient and less strategic and effective approaches.

The European emissions trading scheme is different to the one proposed in Australia in some respects. We would hope the Australian one would be better managed, but the Australian Government is planning to allow purchase of overseas permits or offsets. The European scheme has been based on free permit handouts for existing polluters, based on an assessment of existing emissions, and there are many reports of large scale profiteering from these handouts, as well as maintenance of existing carbon emissions in order to maintain the free handouts.

In one text on carbon emissions trading (45) it is suggested that the carbon trading concept is impractical on a scale of the USA or a global scale. In the USA, hundreds of thousands of industrial sources would potentially need to be monitored. On a world scale, millions of sources potentially need to be monitored. Problems of inaccuracy, exaggeration and cheating are also described.

This text also includes conclusions that "***there is no reason to expect that countries will reduce their greenhouse gas emissions to comply with quotas that cannot be effectively monitored and enforced***' and "***as in every aspect of carbon trading, the tail of free-market ideology is wagging the dog of science, political common sense, and technical possibility***" and "***because it treats all one-tonne cuts as the same no matter where and how they occur, and results in the cheapest cuts being made first, emissions trading runs the risk of delaying progress in dealing with global warming***".

There is argument, based on the problems discussed above, that emissions trading, as an approach to emissions reductions, is not efficient or effective. There is suggestion that emissions trading encourages nothing in the way of strategic changes, but only short term lowest cost "tweaks". There is suggestion that the world will remain locked into high carbon emissions as much as we are locked into the "QWERTY" keyboard.

The author of this book would question what has been achieved by emissions trading in other countries. Where are the big public renewable energy generation facilities? It seems there are not yet many. The criticisms above come from opponents of emissions trading. It is not easy to tell from afar to what degree the approach is successful and to what extent it is not. However, a 2008 analysis by Stanford University (72, 73) of carbon offset processes found that:

"between a third and two thirds" of emission offsets under the Clean Development Mechanism (CDM)—set up under the Kyoto treaty to encourage emissions reductions in developing nations—do not represent actual emission cuts".

Based on this analysis, if we aim for only 5% emission reductions, we would achieve only around 2.5% reductions. Alternatively, we would need to double targets to get the intended result. If this Stanford analysis is correct, emissions trading is not cost effective.

Emissions trading and other newer proposals in the USA

The USA is in a similar position to Australia, in that their American Clean Energy and Security Act (ACES) which includes emissions trading has been passed by Congress but became stuck in their Senate. It proposes 15% of energy from renewable sources by 2020 combined with 5% of energy savings. It proposes a cut of total emissions of 17% below 2005 levels by 2020 and of 83% by 2050.

The emissions trading proposal for the USA is that 85% of permits will be allocated for free and the remaining 15% will be auctioned.

Critics of emissions trading in the USA, including environmental groups such as Friends of the Earth (64), make similar criticisms of emissions trading as those made of the European scheme described in the previous chapter, in that emissions trading is not efficient and seeks out low cost and ineffective approaches. Other interesting criticisms that are made of proposed emissions trading approaches in the USA include:

- That emissions trading is leading to the creation of financial derivatives and other types of middle men that increase the costs of real emission reductions compared to any direct funding of such initiatives. It is suggested that this is evident from the number of carbon trading lobbyists now out-numbering environmental lobbyists by a factor of about 7 to 1.
- That the US markets will be unable to properly regulate the carbon markets and that a "sub-prime" carbon market will evolve that will undermine the market as a whole

- That if the price of carbon were to plummet for any market-driven reason, that will severely undermine both existing and new investment into genuine and effective climate change mitigation efforts.
- That the USA should not allow purchase of carbon "offsets", where the majority of "junk" carbon exists.

Interestingly, in the USA there are also additional proposed bills for "Safe Markets Development" that echo the Australian Greens Party proposed "Safe Climate" bills (described in the next chapter) to ensure the quality of permits or offsets and help reduce these kinds of problems.

Also, very interestingly, the proposed budgets contain substantial commitments of revenue from the emissions trading programme to not only compensating households but also to fund renewable energy projects (65). Of $645B of revenue over 10 years, starting in 2012, the bill mandates investment of $90B in new investments in clean energy, $60B to carbon capture and sequestration, $20B to advanced technology vehicles and $20B to scientific research and development. However the proposed cap and trade scheme may not proceed in the USA.

In the last weeks of April 2010, at the same time as Australia delayed the Australian emission trading proposal, US Democrat Senator John Kerry, in conjunction with a Republican and independent senator, was ready to announce a new, alternative climate and energy bill that has the combined purpose of addressing climate change and also increasing the USA's energy independence. As well as promoting renewable energy, it also promotes nuclear energy, "clean" coal, and increased offshore oil drilling. It requires industry to pay for emission permits but is not a full emissions trading proposal. At the last minute the Republican Senator withdrew his support for the bill, citing a preference of the US Congress to address immigration issues first as the problem. It is reported that the US Republican party in general does not support the bill. This might be because it requires industry to pay for emission permits. Environmentalists are not keen on the bill due to its emphasis on "clean" coal, nuclear and offshore oil based approaches.

The final week of April 2010 was thus a dramatic week for climate change policy in both Australia and the USA. It included the announced delay of the Australian Government's proposed emissions trading scheme. It also included the delayed announcement of a new replacement bill in the USA. It was the start of a major offshore oil spill in the Gulf of Mexico. This all provided a dramatic and unsettling start to 2010, which is first year of the decade in which we need to start reducing emissions.

Would Australia's emissions trading scheme be successful?

The Australian Treasury papers describe the planned emissions trajectories and mechanisms for industry to purchase permits. It is clear from the principles and theory of emissions trading that the planned emissions reductions would be achieved, at least on paper.

The Australian Greens party appears to have decided to reluctantly support emissions trading in general but is suggesting stronger emission reduction targets are required (40% by 2020) and has opposed the range of industry protections that are proposed as well as other details of the proposal that will reduce its effectiveness (49, 52).

The Australian Greens party, via a set of **Safe Climate** bills (52), thus proposes a set of emission reduction measures that include changes to strengthen the Government's proposed emissions trading scheme, including fixing the price of carbon in initial years to around $20 per tonne, reducing subsidies, setting increased emissions targets, and purchasing "Gold Standard" carbon emission offsets that "ensure emissions reductions are genuinely additional and permanent".

It appears that the proposed amendments from the Australian Greens party would increase the effectiveness of the Australian emissions trading scheme, and help ensure that emission offsets are genuine. The author expects that the down side of such proposals is that they will further increase costs, and so the Australian Government has not agreed to these changes. The author has asked the Department of Climate Change to provide some further description of its proposed quality standards.

The Australian Garnaut report (66) which recommended an emissions trading approach, does also say that there are a number of circumstances where an emissions trading approach would not be the best approach, and where a carbon tax would be better. Such circumstances include excessive need for industry subsidies, carbon price limits and types of lack of readiness in the market. Most of the types of circumstances described appear to exist.

Some issues in the Australian emissions trading proposal

As suggested in the preceding chapters, an emissions trading scheme can be ineffective if it is not well designed.

The EU scheme is based largely on free permit handouts to countries and large emitters based on pre-existing emission levels. This provides a baseline from which emission reductions can be achieved. Especially where allocations are over-estimated, this results in profiteering, rorting and cheating, and limits real climate change action when emitters choose to maintain emissions in order to get more free permits.

The Australian system is likely to be better designed but is likely to still have some anomalies.

One that has been pointed out is that the scheme makes it difficult for areas outside of the system to contribute to emission reductions. For example, if households reduce emissions, electricity companies then need to buy less permits, and so more permits are available for other emitters and they become cheaper. It seems to the author though that household reductions would however allow the government to reduce emission caps faster in future years, so they are not totally wasted. In general however, emission reductions cannot occur any faster than the cap set under the target reduction trajectory, and this is another downside of emissions trading.

The system will work well if there are no inaccuracies, where emitters do not understate emissions, and offset arrangements do not overstate emission reductions. That way, for each extra tonne that is emitted, the purchase of an additional offset will reduce emissions back to the cap. That is how the scheme is meant to work in theory, but it requires a high level of accuracy and quality management to ensure that emission reductions are real and not just on paper.

In the author's view the biggest downfall of the proposed scheme, and emissions trading in general, results from how funds from emission permits are used. Especially in the first years of the proposed scheme in Australia, the funds raised from these auctions at $13B are, by a very large extent, the major source of potential climate change fighting funds. They cover the cost 75% of our emissions, being the scope of the scheme. If offsets were purchased at the same price as the permits, Australia could reduce its emissions immediately by 75%.

The Garnaut report (66) suggests that in the theory of emissions trading, the pool of funds from permit auctions should not be used directly to fight climate change, as that would undermine the basic market mechanisms. Under the emissions trading theory that is described, such funds should be given back to households, used for tax cuts, or perhaps used for climate change R&D, which is one of the identified areas where the carbon market "fails".

By comparison, under the proposed US Clean Energy bill usage a good proportion of the funds from permit auctions for direct-action climate change mitigation was mandated.

The following is a discussion about how Australian permit funds will be used, at least in initial years, from the Australian Institute of Public Affairs (58):

> *"The ETS is going to bring the government a huge amount of additional money as it auctions permits each year. Conservatively, that amount will be at least $5 billion extra in the first year, not including the extra GST applied to permits traded in the marketplace. Reducing emissions by 60 per cent of their 2000 levels by 2050-as the government has committed to do-will require making fewer and fewer permits available, and the amount of money the ETS brings the government will increase massively.*

The government has described the ETS as 'revenue neutral', however, the government only means that it plans to spend all the extra money it receives. This is no different from money raised from other taxes. The government has explicitly committed to use much of the proceeds of the tax (the figure of 50 per cent has been mentioned) to compensate low and middle income households, pensioners and other welfare recipients.

Other money will be directed towards subsidising energy efficiency technologies at the household level. In short, the government plans to return much of the money it taxed-a highly inefficient government practice that has been widely criticised in the context of welfare and family policy.

But instead of this elaborate tax churning, the introduction of the ETS should be an opportunity to cut taxes in other areas of the economy. With just the first year ETS windfall of $5 billion, the government could increase the tax-free threshold from $6,000 to $9,000, or entirely eliminate the top 45 per cent income tax bracket."

Thus the major funds at our disposal for reducing emissions would be returned as subsidies, and the emission reduction trading that is done by industry is unfortunately only a comparatively minor undertaking. According to the trajectories being set under the CPRS-5 scenario, these emissions reductions are only around 1% per year, which is really of little effect.

The alternative policy shown later in this book aims to show that we could achieve a much better result if we used these funds in a coordinated manner to fight climate change. Given the size of the funds pool, that is not all surprising. Later in this book a much more aggressive climate change mitigation plan that could be undertaken using these funds is presented. In additional, different funding approaches are described that would ease costs to Australian industry.

A carbon tax?

Following Australia's 2010 federal election, a carbon tax is now very much under discussion. Emissions trading fixes a planned emissions trajectory and allows the price of carbon to vary dynamically according to market demands. A carbon tax fixes the price of carbon, at any point in time, but doesn't pretend to guarantee any particular emissions trajectory.

The web site *www.carbontax.org* advocates a carbon tax for the USA. It proposes an aggressive emissions reduction plan based on a carbon tax. The detail is in a complex spreadsheet that can be opened from the site. In summary it suggests the following plan as an example of carbon pricing and the emissions reduction trajectory it might achieve.

USA

Year	*2012*	*2015*	*2020*	*2025*	*2030*
Carbon price	*$12*	*$50*	*$112*	*$175*	*$225*
Revenue pa	*$68B*	*$250B*	*$476B*	*$634B*	*$811B*
Emissions reduction	*3%*	*13%*	*29%*	*41%*	*$49%*

A quick translation of the suggested model for Australia is achieved by dividing revenue by 20.

Australia

Year	*2012*	*2015*	*2020*	*2025*	*2030*
Carbon price	*$12*	*$50*	*$112*	*$175*	*$225*
Revenue pa	*$3B*	*$12B*	*$23B*	*$31B*	*$40B*
Emissions reduction	*3%*	*13%*	*29%*	*41%*	*$49%*

Such a significant increase in carbon price is not something politicians might want to consider or talk about much, but we'll see. Some Scandinavian countries already have a carbon price over $100/tonne. There is some commentary that suggests investment in renewable energy will not be significant until the carbon price rises to fairly high levels. Some opponents of a carbon tax thus suggest it will be an ineffective approach unless the carbon price gets high enough.

The suggested USA plan is for a revenue-neutral tax where all the revenue is given back to the public as a per-capita dividend. As an alternative approach, trade-affected industries might also be subsidised, which then reduces their incentive for action. All the required climate change mitigation is thus left to industry and consumers, encouraged as a means of avoiding the tax or its flow-on costs.

Whilst no firm carbon tax policy is on the table, it is not possible to be sure of its likely effect. Questions that need to be asked about any carbon tax proposals are therefore:

> *What will be the planned carbon tax price and planned emissions reductions over time?*
>
> *Will government take any direct-action using the revenue or will it be used for subsidies?*
>
> *To what degree does the modeling assume (or can we know) that a tax will result in action?*

This last question is very important. It is about the extent that we expect the tax will result in action or to what degree will the higher prices just be accepted without any action? Can anyone truly model or predict this kind of human behaviour with any certainty? Economists talk about this effect in terms of "price elasticity". Prices of goods such as electricity and fuel are said to be "elastic" if prices go up without impacting consumer behaviour. Usage of one litre of gasoline/petrol produces around 2.3kg of CO2, so a carbon price of $50 per tonne should add only about 12 cents to the pump price per litre. Would this level of price rise have much behavioral effect?

The spreadsheet model summarised above makes arbitrary estimates of the likely effects, in terms of gradual annual emissions reductions, for every $10 of tax. Other economic models might assume industry or consumers will always take the cheapest path. In practice, the real outcomes are probably not known. Industry and consumers might just accept the price increases, and take little climate mitigation action, especially if the carbon price is too low or subsidies are too high.

There is only so much that consumers can do. In the end it is industry and/or government that will need to take the major investment decisions to bring about effective action. What else could prevent action being taken by industry? All the" *non-price barriers"* to action by industry described in the *ClimateWorks Australia* report on carbon pricing (71) will apply. These include:

- *Doubt caused by any lack of clarity of policy*
- *Time delays in decision making*
- *Difficulties in coordination of stakeholders*
- *Splitting of incentives across different parties*
- *Lack of awareness of solutions*
- *Conflicting business investment priorities*
- *Capital investment constraints*
- *Long payback periods and*
- *Uncertainty of carbon price.*

Given both "price elasticity", and the "non-price barriers" to action, a carbon price of $20 per tonne might only encourage $5 worth of action, and a price of $100 per tonne might only encourage $25 worth of action, on average. The principle of a carbon tax is to cause pricing-pain and thus encourage action. However the amount of action taken will generally be less than the amount of pain caused.

From that perspective, it isn't clear whether a carbon tax is an efficient or effective method to achieve climate change mitigation, as some economists claim. Also, leaving all action to industry and consumers would be a lazy approach by government.

Further questions we should ask are therefore:

> *When economists suggest that a carbon price is the most economically efficient mechanism for climate change action, do they and their modeling take into account the types of "non-price barriers" to action by industry that are described here?*
>
> *What mechanisms will be established to ensure the intended level of action actually occurs?*
>
> *Given the enormous risks of climate change, what is the opportunity cost of in-action?*

Given the enormous risks of climate change, the opportunity costs of in-action cannot be easily quantified. Thus there are significant risks associated with a carbon-tax approach, where it is not guaranteed that the intended level of action will actually occur.
A carbon tax, alone, risks in-action, particularly if the carbon price is not high enough to overcome not only the price barriers to action but also the non-price barriers. Climate change mitigation is not the core business of industry. Rather than being revenue-neutral, a carbon tax could be used to fund a large-scale direct-action policy such as the one described later in this book.

A summary comparison of the proposed Australian policies

Based on the potential problems with emissions trading described in the last chapters, as well as arguments against the Opposition's proposed soil abatement approach, this author suspects that neither of the proposals originally put on the table by the Australian political parties would actually achieve their stated emission reduction goals, unless they were modified.

One might ask why the Opposition presents its policy as so much cheaper than the Government's, at only about one tenth the cost, where they both aim to achieve the same 5% reduction in emissions by 2020. As already described the cost comparison is not straightforward.

The Oppositions policy cost is just the cost of the planned emissions reduction, mainly at cheap rates. In comparison, the Government's apparent policy cost starting in early years at around $13B pa. is the cost of permits at 75% of our total emissions, being the scope of the proposed emissions trading scheme.

However since the emissions permit auction funds are not used to fight climate change, the Government will argue they are just circulated back into the economy via various means, and they are not costs at all. They are however the price paid by industry, but not the costs of fighting climate change.

If the permit funds were not used to fight climate change, the Government's policy is therefore actually very similar to the Opposition's, at least in the short term, in terms of scale – just an approach of funding very conservative levels of emission reductions using a collection of least cost options. One is lead by Government and the other is lead by industry.

However, there are still some distinct differences. A summary comparison of the two originally proposed policies is as follows.

Of the two policies originally on offer, the Government proposed emissions trading approach was the policy which can better scale to levels required to fight climate change, especially if the funds from permits were used for that purpose. However, they probably would not be. It had the advantage that it puts a price on carbon emissions and this encourages industry to act. It had a number of disadvantages, being:

- The carbon price is not known, because of potential change to Australia's post Copenhagen commitments, because permits are auctioned, and prices are expected to go up over time
- Up to 50% of investment in new green technology and industry is likely to go overseas
- Some existing industry activity might move overseas ("carbon leakage")
- For industry, it has significant costs

- The pressure on industry is likely to make some areas of Australian industry less competitive and could cause job losses
- It requires significant administration
- Because of the various costs, complexities and uncertainties, it is likely to be unpopular with voters
- The funds from sale of permits will probably not be used for climate change mitigation.

The Australian Opposition's proposal is presented as a much lower cost programme of direct action, of only one tenth of the apparent cost of the Government's proposed policy over the first four years. It is thought to match the Government's proposed target of a 5% emissions reduction by 2020. All investment remains in Australia and it puts little pressure on business and industries. It has the following disadvantages:

- The example plan suggests only a conservative amount of funding to reduce actual carbon emissions.
- Its apparent low cost relies significantly on a low cost soil abatement strategy to sequester carbon in the ground, and there is debate whether this approach can work, or can be measured.
- It has no stated method to scale to high levels of commitment to meet Copenhagen targets or into a longer term strategy.

Post the 2010 federal election, a carbon tax is now high on the list of options being discussed for Australia. It might be used as a stepping stone to a fuller "market mechanism" such as emissions trading at a later stage. Proponents of a carbon tax suggest that has the following advantages:

- It encourages industry to act.
- It is efficient and effective.
- Compared to emissions trading it provides certainty of carbon price that will encourage investment.
- It is simpler than emissions trading.

Opponents of a carbon tax argue that:

- If the price is too low or subsidies are too high, it will have little effect.
- It disadvantages local industry.
- Compared to cap and trade systems, it doesn't guarantee any particular emissions reduction trajectory.

Until details of a carbon tax policy are known, it is difficult to say whether it will be effective, or just a tooth-less tiger that will not achieve a meaningful level of action. If the price rises only very gradually over time, it could be years or decades before any real action is taken.

In the following chapters, a different policy is suggested. Like the Opposition's proposed policy, it is a direct-action policy, but at a much larger scale, at the apparent cost of Australian Government's originally proposed emissions trading programme. It has advantages over emissions trading such as certainty of cost. It proposes funding approaches that requires less of the cost to be borne directly by Australia's industry. It provides more certainty of action than a carbon tax, and possibly more efficiently. Most importantly, it would provide Australians with a very aggressive climate change mitigation policy, but at a cost which is similar to or lower than the Government's previous proposals.

Is there a better policy for Australia?

Introduction

In the author's view, neither of the originally proposed Australian policies were much good. Neither provided a high level of confidence we could limit warming to 2 degrees, and avoid "dangerous" levels of climate change. This chapter of the book puts forward some different policy ideas, based around a programme of large-scale direct action. This is very different to the emissions trading approach proposed by the current Australian Government. It is like the Australian Opposition's proposed approach, but at a much larger scale and with stronger long term goals.

In following chapters we look at different ways to raise the funds and to use the funds.

We look firstly at a different way of raising the funds, because that is the simplest discussion. A simple approach is put forward initially, of raising funds through electricity usage. Other possible methods could include a carbon tax, or a mining tax on coal, oil and gas, or even borrowing.

Raising funds through electricity usage?

In the initial years of Australia's emissions trading proposal, the auctions of permits will collectively raise around $13B per annum. This is for the minimum CPRS-5 scenario. This cost will be shared by approximately 1,000 emitters. That is an average cost of $13M per emitter, which is a large cost. No wonder these companies are complaining and suggesting job losses will occur. There is also uncertainty about pricing to deal with.

Is there a simpler and more effective way of raising such funds for fighting climate that does not involve having the complexities and uncertainties of an emissions trading scheme? There are a number of ways. One good way would be to have a flat carbon tax. Another would be to raise funds through other types of taxes.

One possibly simpler and more equitable way to raise climate change fighting revenue is through electricity usage. For example, if we divide this **$13B** figure by the amount of electricity currently used each year in Australia we get:

$$\frac{\$13B}{260\ TWh} = \frac{\$1}{20\ KWh} = \textbf{5 cents per kilowatt hour}$$

Thus, if we charged all Australian businesses and households an extra 5 cents per kilowatt hour, we would raise $13B per year. This is about double the level of electricity price rise forecast under some levels of emissions trading.

One estimate of average household usage in Australia is 7,000KWh p.a., making the above cost $350 p.a. per household for direct electricity use. This is similar to the estimated cost per household of the proposed emissions trading scheme. In Australia, it equates to one or two of the typical interest rate rises that we have become accustomed to.

Such a method of revenue raising has some similarities to paying for carbon permits, or a carbon tax, but revenue is raised only from electricity usage, rather than other activities. It is simple to understand and a predictable cost for households and industry, as they know their electricity requirements. By comparison, under an emissions trading scheme, the costs of permits is uncertain, as is the cost to households.

The direct cost to businesses and industry will be less than in emissions trading, because a higher proportion of the cost will be borne directly by households. Some tuning of this proposal might be required, with a lower rate for some very big electricity users.

The proposed emissions trading approach could be seen to have a broader base than this approach, in that it covers 75% of emissions, rather than just electricity usage. From a different perspective, this approach could be seen to have a broader base than sale of emissions permits, even though electricity is the source of only approximately 50% of Australia's emissions, because everyone uses electricity.

We will revisit different funding approaches later in the book, but now let's look at the more important issue of how we could use the funds under an aggressive direct-action policy, starting with implementing 100% renewable energy.

Can we afford 100% renewable energy?

Electrical energy generation has risen to about 50% of Australia's carbon emissions, and also comprises approximately 50% of our likely carbon emission savings (38).

As discussed previously, an Australian Government report on the potential for renewable energy in Australia (30) concludes that:

> **"there are no technical or financial impediments to renewables providing about 50% of all Australian electricity demand by 2040"**
>
> and also
>
> **"In the longer term, it appears there is no technical reason to stop renewable energy from supplying 100% of grid energy".**

If there are no technical or financial impediments to achieving a 50% target, and no technical reasons to prevent a longer term 100% target, what would it actually cost for Australia to achieve 100% renewable energy or close to it?

This is explored in the following chapters. Analysis is made primarily for Australia. Attachments to this book contain similar analyses for the USA, China and then the whole world.

Capital costs of renewable energy

In analysing impacts on the electricity market of emissions trading, the consultants to the Australian Treasury used in their analysis (43) a range of current capital costs for renewable energy as follows, in $ per KW.

Wind	$2,134
Biomass - Steam	$2,598
Biomass -Gasification	$2,784
Geothermal – Hot Dry Rocks	$4,200
Concentrated Solar thermal plant	$4,176
Concentrating Photovoltaics	$4,640
Hydroelectric power	$2,320

These costs are generally around $2,500 per KW, but over $4,000 per KW for solar thermal, solar photovoltaic and geothermal facilities. These prices appear to be based on 2008 US Energy Information Administration data.

Remarkably, the electricity impact modeling also lists an assumption that costs of these new technologies will reduce only by around 1% to 1.5% per year.

There are reported price decreases in solar cell technology of around 40% just in past 2-3 years, since this emissions trading modeling was done, and drops in solar thermal technology are starting to occur and more are forecast in the next few years.

For solar photovoltaics one industry analyst report from 2008 (50) suggests that reductions of 7% per annum are expected from 2007 to 2012.

Another analyst report (52) says " *2010 will benefit from sharply lower prices, expected to reach $1.50/W, down from over $4/W in Q3'08, and growing stimulus/subsidy plans in Australia, China, Greece, Japan and the US.*"

There are some reports of solar capital costs very soon reaching as low as $1 per watt, or $1,000 per KW. This based on new thin-film technology that uses less or no silicon.

Thus it appears that parts of the cost modeling assumptions behind the Government's emissions trading proposal of what can be achieved by renewable energy have been vastly out-stripped by recent price decreases and more expected in the near future.

The Victorian Government's planned large scale solar facility (which has not proceeded) was planned to cost $420M for a 154 MW facility. This was planned a few years ago, at approximately $2700 per KW, still much less than the Australian Treasury assumptions. Given that prices of photovoltaic cells are reported to have dropped by 40% in the last 2-3 years, this would potentially bring the Victorian Government facility cost price down to the cost assumption in the Government's Solar Flagships budget.

The Solar Flagships project has a budget of approximately $1.6B for a 1000 MW facility, which is $1,600 per KW. This is much less than modeling assumption in the Australian Treasury documents, but unfortunately prices of solar thermal facilities are probably not quite that low as yet.

Tenders are being evaluated for the Solar Flagship project, so the Government now knows the proposed costs per KW. One company is advertising on the web that it is planning build a 250MW solar thermal plant in Australia for $1B. This is still $4,000 per KW, but commentary suggest this price for large solar thermal facilities could drop by 50% within a few years, once a few of such facilities are built, construction methods are streamlined and R&D costs are recouped.

For the calculations in this book we will thus use a figure of $1,500 per KW, similar to the budgeted cost of the Solar Flagships programme budget, and in line with some of the industry analyst reports, but much less than the $4,640 per KW figure in the Australian Treasury model figures (43). It is in line with where the cost of solar photovoltaics appears to be now, and where solar thermal facilities appear likely to be in just a few years.

Not surprisingly, usage of this lower capital cost assumption predicts that more can be achieved with renewable energy than predicted by the Australian Government's emissions trading models.

Calculating the costs of 100% renewable energy

In the next chapter we calculate the cost of achieving 100% renewable energy in 30 years. This is a much more aggressive target than 20% by 2020. The calculation is under taken not only for Australia, but also the USA, China and the whole world. These latter calculations are shown in attachments.

The calculation is based on current electricity demands, except for China where usage is growing rapidly. For China, the calculation uses China's expected energy requirements in 2030. For the sake of this simple analysis, it is assumed that growth in electricity demand over time will be met by reduced costs as technology matures. Electricity usage in Australia through to 2050 is expected to double. An estimate that prices will only halve in that time is very conservative.

Electricity units you will need to understand these calculations are:

1 TW (terrawatt) = 1000 GW (gigawatt)

1 GW (gigawatt) = 1000 MW (megawatt)

1 MW (megawatt) = 1000 KW (kilowatt)

1 KW (kilowatt) = 1000 W (watt).

1 KWh (kilowatt hour) = 1 kilowatt of power for 1 hour

The analysis is based on an estimate of today's or near future costs of building solar energy plants. In practice, a range of renewable energy sources could be used, not just solar facilities. These will include solar thermal, wind, geothermal and biomass.

Solar facilities have a particular energy output pattern. It is assumed that a solar facility will generate 1500 hours of electricity per annum, being 4 hours a day on average for 365 days per year. This would mean a 1 MW facility would generate 1.5 GWh p.a. A 1,000 MW facility would generate 1,500 GWh p.a. It is a multiplication by roughly 1.5 and then 1000 to convert power to output. In practice this relationship will vary from facility to facility and location to location. Some facilities report much higher output than this, and so this estimate is conservative, but makes allowance for transmission line losses, which can be significant.

The order-of-magnitude calculations in this book are deliberately kept simple. The Australian Treasury models provide more detailed analysis, which encompass a range of different renewable energy capital costs, usage growth, price reductions, transmission line losses, net present values, running costs amongst other factors. Obviously a more detailed and expert analysis of proposals presented in this book could be conducted.

The analysis in this book ignores running costs. This would be valid if it could be assumed that running costs for renewable energy would be similar to current running costs. Solar facilities do not pay for fuel.

The following calculations are thus done on the basis of how many 1 GW "Solar Flagship" project sized facilities (1500 GWh p.a.) we would we need to meet a 100% renewable energy target.

How much would 100% renewable energy cost Australia?

Australia's current electricity usage is 257 TWh or 257,000 GWh per annum.

Using solar only, this would require 170 1GW facilities, as 257,000 GWh / 1500 GWh = 170.

At $1.5B each this is $255B, or $8.5B per annum over 30 years.
Australia's population is approximately 20M. The above cost is $425 per person per year ($8,500M/20M).

Australia's annual gross domestic product is about A$1,150B. The above cost is 0.7% of GDP p.a. ($8.5B/$1,150B), which is less than one typical year of GDP growth.

The above calculations use today's prices, which will reduce. In Australia, a continuing drop in renewable energy prices will probably exceed electricity usage growth, making the above probably an over-estimate of overall cost.

The cost per person above is an average. Depending on how it is funded, those that consume more would pay more, and those that consume less would pay less.

As attachments, similar calculations are made for the USA, China and the world as a whole. For the USA and the world as a whole, the cost of 100% renewable energy calculates to 0.9% of GDP per annum. For China it calculates higher, but China's annual GDP growth is also currently higher.

Based on this, the cost of achieving 100% renewable energy in the next 30 years is well within our financial capabilities. For Australia, the cost is around $8.5B per year, which is less than the total funds we plan to raise to climate change mitigation ($13B p.a.) under this book's policy proposal.

However, electricity generation causes only 50% of our emissions. In the next chapters, we explore how we might make further emission reductions to get to carbon neutral.

Could we be carbon neutral, or better, in 30 years?

Climate change analysts suggest that much more aggressive targets that those currently being set will be required to avoid "dangerous" levels of climate change. To do this the IPCC suggests world emissions need to start reducing by 2015, in 5 years time. Cuts of at least 80-90% by 2050 will then be required to stabilise atmospheric CO2-e levels at 450 ppm or less in order to have an estimated 50% chance of limit warming to 2 degrees or less. Could we construct ourselves a better than 50% chance by constructing a more aggressive policy?

We have calculated that in Australia complete replacement of our electricity supply requires around $8.5B per year over 30 years. The total proposed annual funding of $13B p.a. therefore allows for not only complete replacement of our electricity supply with renewable energy, but also around $4.5B of annual funding that could be used for emission reduction strategies in other required areas.

The author suggests that over the next 30 years the capital cost of renewable energy will come down at a greater rate than our electricity usage will go up, meaning that out of the total $13B per annum budget we would have at least $4.5B each year in today's value to put towards reducing emissions in other areas.

Replacement of our electricity supply with renewable energy is a massive project, but unfortunately is the easiest area to achieve large-scale emission reductions. After that, it gets more difficult, as we will see in following chapters, where we explore how a policy could be developed to eliminate all the other remaining emissions.

Australia's emissions are forecast to approximately double by 2050, to around 1200 MT p.a. if we were to do nothing.

However, there will be a wide range of low emission technology advances by 2050 which are difficult at this stage to accurately predict and quantify. We will be driving more electric cars, and hydrogen refueling stations will start appearing in cities for other types of cars. We will be growing more trees and carrying out less deforestation, out of concern for the environment. Farmers will be carrying out more carbon neutral farming practices. Australia will perhaps be building carbon neutral cars and selling them to the rest of the world under strict new rules that have severely restricted building of petrol engines. Airbus is designing hydrogen fuelled planes. The hydrogen will be made at airports.

To accelerate these initiatives and get us to carbon neutral by 2040, our remaining $4.5B per year could be spread over each other emission reduction area, thus giving a number of substantial funding packages to each area over 30 years. These other emission reduction areas include transport, forestry, land-use, agriculture and waste management, as well as other new approaches to reducing carbon in the atmosphere.

A draft large-scale direct-action climate change mitigation policy for Australia

The following is a draft 30 year expenditure plan, as an example large-scale direct-action climate change mitigation policy for Australia. At $13B p.a. the total expenditure over 30 years is $390B in total in today's dollars. It targets selected types of emission reduction activities. It is written a bit like a government policy document.

$255B	**Complete replacement of electricity supply with renewable energy**. This major capital project creates tens of thousands of jobs. Gas fired power stations may be maintained as backup power. This will save about 50% of Australia's emissions, now and into the future, currently at nearly 300MT per annum.
$10B	**Reforestation of Australia.** At $10 per tree this is 1 billion trees or 50 trees for every Australian. These great eucalypts will create a carbon sink over their growing lifetime and will contribute greatly to Australia being net carbon neutral. This approach is very effective. Reference 62 suggests it takes 300 trees to cancel out a person's lifetime emissions. More analysis of the potential of this approach is discussed in the Australian Garnaut report (66). Under this policy a saving of 50 MT per annum over existing forests will be targeted.
$5B	**Soil abatement trial.** This approach will be attempted, with a higher initial budget than the Australian Opposition party's proposal, and with the aim of achieving a lower target of 100 MT of abatement pa. A number of different soil abatement approaches will be trialed.
$8B	**CO2 capture research.** This substantial budget will be used to fund research into clean approaches to reducing CO2 in the atmosphere, with the aim of developing a commercially viable approach for world-wide application, and in the hope of claiming Richard Branson's $25M prize (70). The substantial budget for this activity anticipates and includes the building of the first large scale commercial facilities.
$10B	Additional agricultural assistance, in a form to be determined, to help reduce emissions in that area.
$7B	Assist development of **hydrogen and electric green cars** in Australia.
$10B	Assist development of **hydrogen refueling facilities** for green cars.
$75B	**Green car subsidy.** There are 15M cars in Australia. Purchase of 5 million new green cars built in Australia will be subsidised for Australian new car buyers. The average subsidy over 30 years will be $15,000 each, but starting at $30,000 each in year 1 and reducing over 30 years. Initially a $45,000 electric car will be available for $15,000, and a more advanced and powerful $75,000 car will be available for $45,000, and a $120,000 hydrogen luxury car will be available for $90,000. $75B will subsidise 5 million cars. There will be a waiting list for these cars. 50,000 of the cheaper cars will be used as public cars in major towns and cities. After the subsidies cease in 30 years only green cars will be available for sale in Australia. Along with other improvements to transport efficiency, a saving of 50MT per annum is targeted.
$10B	**The icing on the cake.** As part of the new world Accord, Australia, with contributions from the USA, China and other participating countries will **buy out deforestation activities in the Amazon** and other major world rainforests, thus saving an estimated 20% of the world's net carbon emissions. Taking credit for 1/3 of this action would put Australia at around 500% net carbon negative. As a more targeted option, working in Partnership with Indonesia only takes Australia to 180% carbon negative. Details are discussed further in a later chapter and in the Garnaut and Stern climate change reviews.
$390B	**Total funds required over 30 years**

Our 5 cents per kilowatt hour goes a surprising very long way, if spent effectively.

In summary, the plan includes 100% renewable energy, planting a lot of trees and accelerating the adoption of electric and hydrogen cars. It trials soil abatement and other forms of agricultural carbon abatement. Combined with other countries we stop deforestation in the Amazon and other countries.

It augments and accelerates other emission reduction initiatives that will occur over the same time period, such as increased building efficiencies, hydrogen fuelled passenger aircraft, and improvements in waste management.

This kind of direct action approach can co-exist alongside emissions trading, if the pool of permits funding were directed towards such scale of direct action.

The emissions saving targets under this proposed policy are thus summarised as follows, being proportions of Australia's current emissions.

Electricity	over 250 MT pa
Soil abatement	at least 100 MT pa from this and other improved agricultural practices including "carbon farming"
Forestry	at least 50 Mt pa, from accelerated onshore reforestation
Transport	growing to 50 MT pa, based on expected advancements and with acceleration from proposed electric car replacement subsidies
Other efficiencies	50 Mt pa. These will be achieved across a range of sectors, including increased building efficiencies, and emission reductions in waste management and mining.

This totals 500 Mt per annum, which is very close to carbon neutral onshore based on Australia's current emissions of around 560 Mt p.a. It is possible that with more thorough analysis of improved agriculture, soil and forestry-based sequestation approaches, we could achieve better than this.

Adding on benefits of rain forest reforestation off-shore, say in Indonesia, takes Australia to about 180% carbon negative (50% credit as a result of equal partnership with Indonesia, of three quarters of 5 times of Australia's current emissions – see later discussion).

More discussion of the proposed approaches as well as other approaches is provided in following chapters.

Tree planting

The emission reduction benefits of planting trees are difficult to quantify. It is beneficial, and probably very cost effective, but the value is assessed differently in different estimates. Some suggest an acre of trees is required to replace 10,000 km of car travel. That will require a lot of acres. Some suggest that when the trees die, replacement trees are then required to maintain the same level of carbon sequestation. From that perspective, it is a strategy of CO2 reduction that can be used once and isn't an ongoing strategy.

If planting lots of trees helps save us this century, then it is a good policy to use to give us time to fix emission levels in other areas. The number of trees we can really plant, and the benefits gained, needs more careful assessment.

The Australian Garnaut report (66) contains a detailed analysis of the potential of this approach. A target of 50Mt per annum savings, as a temporary measure to reduce emissions this century, could be practical.

The author understands that the Kyoto protocol recognizes carbon reduction from reforestation using eucalypts, if that matters.

New green cars

There are 15 million cars registered in Australia (3 for every 4 Australians). If they were all in use, replacing them all only gives is a 45 MT pa emission saving, based on emissions of 3 tonnes of CO2 per annum per car.

Replacing our cars is a small helpful step in saving the extra 300 MT of savings required to be carbon neutral based on today's emission levels. Also, the savings in fuel costs will be massive.

Saving the world's rainforests

The Australian Garnaut review (66) references saving the world's rain forests as an option for global emission reduction. Emissions from global deforestation is thought to contribute nearly 20% of the world's emissions.

The Stern review suggested that US$5B-$10B of expenditure per annum would stop deforestation across eight countries where 70% of deforestation occurs (Cameroon, Congo, Ghana, Bolivia, Brazil, Papua New Guinea, Indonesia and Malaysia). This is a low cost abatement opportunity of $1-$3 per tonne of CO2. Our neighbour Indonesia's total CO2 emissions are thought to be 5 times Australia's, with three quarters of that total arising from deforestation.

Currently, stopping deforestation, and also reforestation, is not encouraged by Kyoto standards for emissions accounting. This is for a number of reasons, including:

- Tree based carbon storage is not permanent
- Results are difficult to quantify
- If deforestation activity is stopped at one location, it can just recommence at another.

The fact that real progress in this major world emissions area is being thwarted by arguments over carbon accounting standards, associated with emissions trading, is an abomination.

The Garnaut review suggested that under these circumstances a real approach to solve this problem would involve Government direct action, ("via a mix of regulatory or fiscal measures"). This approach would obviously be cost –effective. This should be added to the next world accord (discussed in a later chapter).

For Australia, the best approach could be to work in partnership with perhaps two major rainforest countries, with the aim of completely stopping deforestation and commencing reforestation.

What else can we do?

Other major emission areas that we can reduce include waste management, fugitives (escaping gases from coal mines), agriculture (at 16% of emissions) and other forms of transport such as trucking, air and rail.

We can increase energy efficiency, but if we have replaced our energy supply, that is less important.

Some analysts suggest that reduction of other gasses apart from CO2 provides a key area to target. Emissions from waste management are surprisingly high, with some analysts suggesting that emissions from household waste are similar to that of running a car. This an area to address but the author has not researched how much benefit can be gained in this area.

Other analysts, who have obviously looked at this overall emissions reduction problem have concluded that in order to save our climate we will need to employ measures that take existing CO2 out of the atmosphere, as well as reducing emissions. Some more approaches, such as these, are therefore discussed in the next chapters. Some bad approaches are also discussed.

Carbon capture and storage – "clean" coal

This term carbon capture is not used here in a general sense of capturing carbon out of the atmosphere but in the sense of capturing CO2 leaving coal fired power stations – the "clean" coal approach. It is a means of reducing emissions from coal fired power stations as an alternative to renewable energy.

The Australian Government is investing in this approach at present, with more funding from the Clean Energy Initiative being provided in this area than for the Solar Flagships project.

This author has observed the following.

- The Government report on the potential of renewable energy in Australia (30) suggests that renewable energy will be a cheaper approach than "clean" coal in the next 10 years.
- There are risks with underground CO2 storage, with some reports (68) where natural sudden outgassing of CO2 from within the earth has suffocated animal life in the surrounding area, as CO2 is denser than air and sits around the land surface for some time.

In the author's view, the thought of Australia aiming to pump some hundreds of megatonnes of CO2 into the earth does not sound at all "clean". The problem is much greater on a world scale. There may be cleaner approaches to clean coal, such as algae based approaches discussed later, or mineralization based approaches, but the approach of injecting very large amounts of CO2 into the ground seems to be the predominant one.

It is said that commercialization of these approaches is at least 10 years away. It is not thought that we have that long to act.

Transportation of CO2 to storage sites will require CO2 pipelines. The Garnaut report (66) and ZeroCarbonBritain (67) describe the use of CO2 pipelines in the USA as a means for the oil industry to pump CO2 into the ground in order to extract more oil. The author assumes that this technique is the origin of this CO2 sequestation idea.

Some practical problems include how to transport the CO2 to viable storage sites, finding sites that seem sufficiently distant from populations, and the costs of the approach, which would add to existing electricity generation costs, and may exceed costs of renewable energy based approaches.

Expected leakage from storage sites is reported to be low at 1% or less per annum, but this is conjecture and experimental. Opponents of this approach point out that if a significant quantity of stored CO2 was too leak at some point in the future, say over 50 to 100 years, then future generation would find themselves back to square one in terms of atmospheric CO2 content. Such leakage could be uncontrollable.

In the author's view, the practicalities, economics, motives and ethics/morality of the approach need to be urgently questioned. The idea of making coal fired power stations clean is sound, but to the author, the idea of pumping hundreds (or millions on a world-wide scale) of megatonnes of CO2 into the ground does not at all sound "clean" and probably less clean and more dangerous than nuclear energy. More reading about this approach and other possible cleaner approaches is at reference 69.

Coal is Australia's largest export, and Australia is the world's largest exporter, so no doubt there are substantial commercial interests at play.

The amount of money involved in this industry in Australia is large, but is actually very small compared to the size of the world's renewable energy market, as demonstrated in the numbers shown later in this book. The size of the opportunity in the USA alone is massive. Short-sightedness and lack of agility, due to our existing commercial interests, could prevent Australia from taking advantage of this emerging renewable energy market – we have missed a few boats already – as well as hampering our overall climate change mitigation efforts.

The nuclear option

A report on nuclear power capital costs (59) suggests capital costs of around $1,000, $1,400 and $2000 per KW for nuclear power. Given these are comparable to recent capital costs of solar power (see earlier discussion), the nuclear option does not appear necessary for Australia given our large land area and good access to solar energy. The nuclear option may be a good one for other countries with small land areas, less sun light and large populations. Nuclear power is becoming safer than it was.

Soil abatement

As discussed earlier, the feasibility of this approach is debated. It is possibly an approach that could help save us from "dangerous" levels of climate change in the short term.

It is not simply about tilling plant matter into the ground, as microbes will convert this to CO2 in a matter of months. There are however better tilling practices, known as conservation tilling. Some of the better approaches are about growing pasture in the same soil as crops, and promoting more living root material in the soil and contributing to soil humus. There are a variety of approaches. It involves significant changes to current farming approaches.

Advocates of the BIOchar approach, where the plant matter is first burned to charcoal, claim that the additional burning process results in carbon that will stay in the soil for thousands of years.

The Australian Garnaut report (66) provides much detailed discussion of these types of approaches. An interesting summary of the possible approaches from this report is reproduced in an attachment to this book. The tables show that the abatement potential of these approaches is large. The report suggests that costs start around 20 per tonne (as opposed to $10 per tonne suggested by the Australian Opposition party). It breaks down Australia's land area in some detail and explains where the different approaches could be used.

The possibilities for "carbon farming" are also discussed as a way of sequestering carbon. Reduction of many hundreds of megatonnes p.a. of CO2 or more appear to be achievable for a period of some decades. This alone would make Australia carbon neutral. Eventually, some limits would be reached.

As long as these techniques can be used at little expense to ongoing food production, the approaches can be used as a way of achieving carbon abatement over the next 50-100 years to help save us from "dangerous" climate change.

The Kyoto protocol does not fully recognize such approaches, probably because they are difficult to measure, but if they work effectively they are a very good approach. This is another example of how emission trading rules could hold back climate change mitigation, unless we ignore them.

Algae

Algae reactors provide a cleaner approach of extracting CO2 from coal fired power stations. Massive algae farms would be associated with each coal fired power station, and algae converts CO2 to biofuel.

Skeptics of this approach suggest that the cost of building and running these massive facilities, as an additional cost to existing power stations, would currently exceed the cost of a replacement renewable energy facility.

As a geo-engineering process, the idea of putting CO2 consuming algae into our oceans is also being investigated, but this is likely to cause alternative environmental problems.

Other means of atmospheric CO2 removal

Regularly in the news we hear of research into ways of removing CO2 from the atmosphere. Achieving a viable means of doing this on a large scale would have massive advantages. Examination of these proposals appears to show problems of scalability and cost that exceed those of existing solutions such as renewable energy.

Richard Branson has offered a $25M prize (70) for the first organisation to develop a commercially viable option to remove CO2 from the atmosphere. Entries closed on 8 January 2010. As at September 2010, no prize had yet been awarded (70).

Geothermal energy

If the current research into geothermal energy in Australia were successful, this might provide a cheaper approach than solar facilities for electricity generation. This would then leave increased funds to help reduce emissions in other areas.

Agriculture

Agriculture is currently the source of 16% of Australia's total emissions. This becomes an even greater proportion of remaining emissions once we have fixed other areas. For example, it becomes 32% of remaining emissions once 50% are fixed through renewable electricity generation. It appears to be a bigger source of emissions that all our cars.

According to the Garnaut report (66), around 60% of the total agricultural emissions result from methane from cows and a further 20% is methane from sheep. If we need to stop eating beef and lamb in order to save our climate then this author is prepared to do it. Poultry and pork will become the better or cheaper options. Kangaroo and emu also emit less, as they are also not "ruminants", and aren't too bad to eat as an alternative source of red meat.

The Garnaut report (66) contains some very interesting calculations of the number of millions of kangaroos Australia would potentially farm to replace beef and lamb. Australians eat around 10 times as much beef and lamb as the world average. The calculations show that quite a number of kangaroos would be eaten each year by each Australian.

Maybe other countries of the world might also take this up. Kangaroos seem to breed well.

For readers having problems at this point, in its chapter on transforming land-use, the Garnaut report also discusses techniques to reduce emissions from cows and sheep. There is mention of "anti-methanogen" technologies, which include fatty dietary supplements that reduce emissions somewhat. Another approach is switching to "mini cows", which are cleaner.

The Garnaut report suggests however that the likely emission reductions in Australia to be expected through carbon pricing impacts on beef and lamb are only of the order of 16Mt pa. At those modest levels of improvement, the emission reductions from beef and sheep are insignificant compared to the possible hundreds of megatonnes of CO2 abatement that could be achieved this century by other types of improved agricultural practices.

On that basis the author would propose that we keep our cows and sheep for now and focus on carbon abatement in other areas of agricultural emission reductions, such as soil abatement as well as improvements in forestry.

It is argued that an impediment to including agriculture in an emissions trading approach is the many tens or hundreds of thousands of parties involved. If we don't take an emissions trading approach, this avoids much of that problem, but the scale of endeavour and expenditure to change agricultural practices will remain large but worthwhile.

Other policy options and variations

There are potential problems with the draft plan that is presented, which we will call Plan A. We don't yet know for sure the outcomes that can be achieved through large scale forestry and agriculture improvements. The outcomes from forestry might be over-estimated and the outcomes from agriculture might be under-estimated. This gives rise to alternative policy options, which we would need to assess as we move forward. Some examples are as follows.

Plan B – Increased agricultural and forestry action	As per Plan A, but if the agricultural and forestry initiatives prove successful, funding in these areas would be increased, via an increase in overall budget or by reducing the budget for vehicle subsidies.
Plan C – Increased offshore rain forest protection	As per plan A, but if the agricultural and forestry initiatives prove unsuccessful, or limited in their application, increased funding would be directed to offshore rainforest initiatives.
Plan D – Purchase miscellaneous carbon offsets	If neither the onshore agricultural and forestry initiatives, or offshore rainforest initiatives prove successful, we could purchase onshore or offshore offsets, in a uncoordinated, least-cost emissions trading like manner, but in an aim to meet our overall emission reduction targets.
Plan E – Renewable energy only, but faster	Recognising the complexities and research required to obtain emission reductions in other areas, we could embark most aggressively with renewable energy only, in an aim to achieve 100% renewable energy in 15 years rather than 30. This would reduce cumulative emissions in a similar manner to achieving higher emission cuts over a slower time period. A focus on other areas could come later after more research into the certainty of other approaches were undertaken.

Further policy development

The draft plan is by no means a perfect plan. This is only the beginning of a large-scale direct-action climate change policy development. A larger team of experts would provide a more detailed plan and perform ongoing monitoring, review and improvements to the plan as new options were developed. Also, in order to better research, develop and manage the best and most effective approaches forward, task groups to oversee and steer emission reduction activities in each sector would need to be established.

In comparison, the philosophy of the emissions trading approach is that the market will select the least cost options as we move forward. For a direct action approach, strategies must be selected and initiated by a coordinating authority. The reader should decide which is the better approach. Some further guidance on this is provided in following chapters, as well as a suggested conclusion.

The good news is that Australia would not have to act alone. The R&D funding that can be allocated by larger countries such as the USA and China far exceeds what Australia can do alone. Any Australian plan should thus be constructed in concert with these other countries, avoiding duplication of effort and to maximize global results. We do not need to solve green car, atmospheric carbon reduction and world rainforest protection measures alone. Acting with these other nations we could very quickly put in place a new world accord to spend climate change mitigation funds in the most effective manner to aggressively fight climate change.

The Copenhagen Accord agreed to the global allocation of $100B to the purpose of global climate change mitigation. The figures in this book show that in fact we can, and need to, go far beyond that. Australia, in concert with the USA and other nations, could very quickly put in place a much more aspirational policy for renewable energy, rain forest protection, CO_2 sequestation R&D, other R&D and better emission reduction targets that all nations would envy and aspire to. Such efforts will have economic benefits. Rather than leaving all action to industry, this level of coordinated climate change action would be the basis of a new world accord.

A final policy discussion

Funding alternatives revisited

At the start of this policy discussion, this book proposed an alternative funding approach for a large-scale direct-action approach based on funding from electricity usage. It is a funding mechanism that spreads the funding load, is a method the public readily understands (it is easier to see electricity usage than CO2) and would be very easy to administer.

Electricity usage provides a logical and very simple means of raising climate change fighting revenue given that it is the biggest single area of emissions and has greatest potential for emissions reductions. It covers approximately 50% of Australia's emissions, which is very close to the initially proposed coverage of emissions trading, at 75%.

However, a disadvantage of raising climate change fighting revenue through electricity usage alone is that it is does little to encourage emission reduction in some other industry sectors, such as transportation. This does not mean that a Government, through a large-scale direct-action policy, could not use some funds to help bring about emission reductions in these other areas.

A middle ground between that approach and emissions trading would be a **carbon tax**. It is similar in some ways, in that it can provide a fixed price and also would add a fixed amount to electricity charges. The advantage of a carbon tax over an electricity tax is that it provides more incentive to reduce emissions in other areas where industry can readily assist, including agriculture and transportation. Advantages of a carbon tax over emissions trading is that it does not have the complexities and drawbacks of permits, offsets, auctions and trading.

Another hybrid option could be to raise the bulk of revenue through electricity generation, but still introduce a **reduced scope emissions trading scheme**. Thus as well as excluding agriculture, emissions trading could also exclude electricity generation, even though this is actually the largest single source of emissions. The basis for exclusion would be that direct action will be taken to completely replace our electricity generation with renewable sources, which is being separately funded. We could call this new reduced scope emissions trading "**CPRS –lite**". Such an approach could be seen as simpler, smaller, and thus more appealing and more easily understood by voters.

The recently proposed Australian tax increase on mining, particularly on **petroleum and gas,** was originally intended to result in an additional $9B of government revenue p.a. This could have provided another way of taxing emissions, as a possible alternative to emissions trading, and showed there any many ways the levels of funding discussed in this book could be raised. This revenue was however not intended to fund climate change mitigation, and was be used for purposes such as increasing Australian retirement savings. Assuming climate change does not impact our economy, we would be richer in retirement.

ClimateWorks Australia's Low Carbon Growth Plan

There is obviously scope for much more detailed analysis of options for a large scale, direct-action climate change mitigation strategy, or emissions reduction options that are market driven. Under either approach, the best and/or least cost options can be reassessed over time.

There has been another very detailed strategic assessment of potential emission reductions in Australia. In March 2010, ClimateWorks Australia released a **Low Carbon Growth Plan** report (71). The work was based on similar work undertaken by a US-based ClimateWorks Foundation. ClimateWorks Australia is a partnership between the Myer Foundation and Monash University.

This work identifies 249 Mt of reductions that could be achieved by Australia in 2020 with a net cost of $185 per household, or $1.8B in total. This is a 25% emissions reduction from 2000 emission levels. It identifies 54 emission reduction opportunities, including up to 48 Mt from forestry, but only 32 Mt from agriculture. 25Mt of this, only, is from soil abatement. It uses a graphical approach to showing the cost and size of emission reduction strategies.

The full cost of the initiatives is estimated at $7.3B or $749 per household in 2020 but reduces significantly, to the net costs above, due to cost reduction benefits received in areas such as increased building efficiency.

The example plan doesn't suggest the scale of the renewable energy proposal put forward in this book. It suggests by 2020 we might achieve only a mere 3.2 GW of new power generation from solar, and a bit more than this at 4.5 GW coming instead from new gas powered stations. It assumes the cost of solar facilities remains high, thus limiting such work. This is on top of what is achieved by the 20% Renewable energy target, which is regarded as business as usual. It shows more use commencing of "clean" coal and gas-fired energy approaches. The reason for this is that the strategy described is one of least-cost first, with the first approaches taken being of increasing energy efficiencies, with positive payback rather than a cost. Solar energy is by comparison seen as expensive and so it is suggested that little of this is implemented under this least-cost first plan by 2020.

It strongly advocates a carbon price to encourage the proposed investments. The level of emission reductions that is proposed and costed closely matches the more aggressive Garnaut-25 emissions trading scenario in planning a 25% emissions reduction by 2020. It aims to show that this level of emissions reduction is affordable.

However, the report also describes at some length how a number of "non-price barriers" could prevent investors from acting, and achieving the full potential of the emissions reduction plan. These include doubt caused by any lack of clarity of policy, and indirect transaction costs resulting from time delays in decision making, coordination of stakeholders and splitting of incentives across different parties. Other inhibitors include information gaps such as lack of awareness of solutions, any conflicting investment priorities, capital investment constraints and long payback periods.

It suggests how government might help industry overcome these problems, but does explain very well how having a carbon price alone will not necessarily bring about timely action at expected levels from industry. This author would add that lack of certainty of carbon price would be another potential barrier to investment.

Some other important things to note about this ClimateWorks Australia analysis are:

- The report advocates putting a price on carbon. Neither the predicted gross cost or net cost appear to include the total cost of permits, if say an emissions trading scheme were used, or the total cost of a carbon tax. The stated gross and net cost figures appear to only represent the cost of the proposed emission reduction activities.
- The analysis does not seem to include any notion of offsets or permits being purchased overseas, as in Australia's proposed emissions trading approach. The analysis concentrates on analysing the least cost and most efficient onshore mitigation options.
- The analysis suggests that demand for electricity usage would fall by 2020, rather than increase (241 TWh in 2020, compared to 257 TWh now in 2010 or 304 TWh in 2020 based on otherwise expected growth). This is based on the assumed model that a very significant amount of energy usage efficiencies will be implemented by industry ahead of other abatement approaches, because they are thought to be least cost approaches with a positive payback. The analysis therefore seems to this author to represent an idealized, theoretical solution, rather than a probable real-life one. It also seems to assume we don't forge ahead with electric vehicles for example. This demonstrates there can be many different models of our probable future.
- Because the analysis seems to present an idealized solution, and because of the "non-price barriers" to action that are described, the author suspects that the level of mitigation achieved by industry based on carbon pricing alone is probably overstated, and the costs of achieving the target 25% emission reduction outcome are probably understated.
- Importantly, the report concludes that a combination of both carbon pricing and direct action will be required to most effectively reduce emissions.

Why carbon pricing and least-cost first is dangerous

A least-cost-first approach, encouraged by carbon pricing approaches where the carbon price rises gradually over time, is an inefficient, ineffective and dangerous approach for climate change mitigation because of the following reasons:

- It will delay putting in the hard yards until later, and might make only piecemeal improvements in the interim.
- It encourages less clean or less effective approaches in the short term, such as gas-fired power stations, which is then a lost opportunity to do something better, and is thus a less efficient use of funds.
- It might result in a lot of activity being funded which doesn't get us most efficiently and effectively to the end result.
- It might leave addressing the hardest areas, or those with the longest lead times, until it is too late.
- It wastes time, which we have even less of, than money.
- More effective long-term approaches, that are not currently least cost, could become least cost if they were adequately funded. Solar energy is thought to be an example of this.

From the figures we have seen in this book, it is clear that more substantial emission cuts will not be made until electricity is generated with renewable energy. There is a lead time to achieve this, and we should start now, even if this is not currently the least cost approach.

It seems the economists who came up with the theory of emissions trading did not also integrate ideas of project management and critical path analysis. If, for example we had a deadline of 30 years to achieve an objective, and if one of the larger tasks would require 30 years to implement, then it would need to commence immediately, even if it isn't currently the lowest cost activity. Time to implement could be constrained by practical construction timeframes, common sense, or budget.

The theory of emissions trading, as well as carbon pricing approaches such as a carbon tax, also appears to ignore the non-price barriers to action that are described by Climate Works Australia.

The debate comparing least-cost first approaches, based on a carbon-price and industry action, vs a strategically guided direct-action plan is at the heart of the debate around climate change mitigation policies and emissions trading. A least-cost first approach risks not urgently promoting extensive use of renewable energy and assisting to make it the least-cost approach.

A new, revised and optimised method of emissions trading or carbon pricing needs to be established that accommodates higher priced options when a project plan says that they are required. Perhaps when we do a more detailed plan, we will develop this important and essential method.

The principle of least cost first would disappear if a carbon pricing approach began with a realistic carbon price, reflecting the true cost of carbon now, rather than gradually increasing it over time. Based on IPCC estimates, this would be somewhere between $100 and $300 per tonne. This would allow effective climate change mitigation to begin immediately, which is actually what we need.

Getting the balance right

Recently on Australian television President Obama of the USA suggested that it is important to put a price on carbon in order to encourage industry to act, rather than relying totally on government command and control. It is hard to argue with that wisdom.

In this author's view, the most important conclusions from the policy discussion in this book are:

- Putting a price on carbon is important, in order to encourage industry to act, rather than relying totally on government command and control. We should try to ensure that pressure is not excessive, and put local industry at a disadvantage to the rest of the world. This book suggests some alternative ways to better share the funding load.
- Putting a price on carbon alone is not sufficient, as we can't rely totally on industry to act efficiently, strategically and in a coordinated manner, for reasons discussed in previous chapters, and also due to the "non-price barriers" to action described very well by ClimateWorks Australia. Just as much, if not more, can be achieved with effective Government direct-action, command and control policies, as demonstrated in this book.
- In terms of achieving results, the most important thing is not how the money is collected but how it is spent.

So the best policy is brought about by getting the balance right. The right policy will:

- Spread the funding load directly across industry and the community, and not merely to large emitters, making them uncompetitive internationally.
- Achieve a balance between industry-lead action and Government-lead direct- action, and not leave all the required action to industry. Industry leaders primarily have their main core business on their minds.
- Plan and coordinate the mitigation effort in a strategic manner. A least-cost, market-driven, uncoordinated carbon pricing based approach is probably not the best way.

In this author's view, an emissions trading or carbon tax working alone is not the best policy because it doesn't get these balances right. It is however, much better than doing nothing. Contrary to emissions trading theory, the funds from permit auctions could be used for Government-lead, direct action, alongside industry action, resulting in much faster emission reductions.

Alternatively, an emissions trading approach could be replaced by an alternative carbon pricing approach such as a carbon tax. The same arguments to accompany this type of economic lever with strong government-lead action would also apply. A carbon tax in itself does not guarantee effective action.

.

Summary of recommendations to governments

This chapter has been updated post the Australian 2010 federal election, as the topics of debate have now changed significantly. Also, in the context of Australia's new minority Government, and a multi-party Climate Change Committee, it now presents recommendations to Government as a whole, rather than individually to each Australian political party as it did prior to this election.

The same recommendations presented below to the Australian Government would apply to the US Government and other world governments. Presently most governments seem equally confused about how to address the significant problem of climate change.

The author's recommendations to the Australian and/or other governments are:

1. Forget emissions trading for now.	Forget a full cap-and-trade based approach to emissions trading for now. It appears unlikely to be adopted on a world-wide basis in the short term due to its problems and impracticalities. It is subject to too many middle-men, rorting* and inefficiencies. It is too complicated and has too much pricing uncertainty. Some more limited, efficient and effective forms of trading might be adopted at a later stage. * This is an Australian term.
2. For a carbon tax, ensure the price is high enough to limit warming to 2 degrees. Include additional mechanisms to ensure the process works, because it might not.	If considering a carbon tax as the main policy platform, ensure that the carbon price is high enough to encourage a meaningful level of climate change action, based on an effective emissions reduction trajectory that will limit warming to less than 2 degrees. Include an independent monitoring and review process that will ensure an effective level of action occurs, and we achieve the planned trajectory. Because of not only price barriers, but also non-price barriers to action (see ClimateWorks Australia -71), industry might carry out less climate change mitigation than you expect. It is not their core-business.
3. Do not accept that a carbon tax is the most efficient and effective approach, even if some economists and industry leaders tell you so.	Do not blindly accept claims from economists or industry leaders that carbon pricing approaches are the most efficient and effective means of achieving climate change action. These claims do not appear to be presented with a full range of more balanced argument regarding the advantages and disadvantages and likely effects of these policies. Organise an independent, urgent and thorough review of claims that a carbon pricing approach will be efficient and effective. This should analyse not only the price barriers, but also non-price barriers to action.
4. Consider whether a direct-action approach would be more efficient and effective.	Consider a programme of large-scale direct-action, such as the one presented in this book, as an alternative approach to climate change mitigation. Consider its costs, benefits and compare it to the costs, risks and level of reliability of carbon pricing approaches which expect industry to act. Consider funding a programme of direct-action using a carbon tax, feed-in tariffs or other methods described in this book.
5. Avoid least-cost first approaches. This delays action.	Do not plan a programme of climate change mitigation based on least-cost options first, such as energy efficiency, and leaving the harder and more costly options until later. This only delays action. We have limited time to act. If using a carbon tax, ensure the price level is effective from the start.
6. Achieve international agreement.	Turn your sights to the next World Accord, where we will explain and promote the chosen approach to the world, and help set better international emission reduction methods and targets to avoid "dangerous" levels of climate change.

Large scale solar thermal facilities

Are the ideas regarding renewable energy put forward in this book practical? The author thinks so.

Even though the steps the world is taking towards fighting climate change are not yet enough, we are advancing. For 2020, Australia has a 20% renewable energy target, China has announced a 15% target which is a very big commitment, and the USA is looking at a 17% emissions reduction target from 2005 levels.

As renewable energy prices drop, and temperatures gradually warm, world Governments are likely to start setting better targets.

Solar thermal facilities offer great potential to meet these better targets, as do many other options.

If the reader was to Google or Bing "large scale solar thermal facilities", you will see the kinds of facilities that are now being started in California and China and are proposed in Australia.

These facilities are very large, of up to several square kilometers each. The look of our landscape as we fly over it will change. They are a bit ugly, but not as ugly as climate change is forecast to be.

Scientists suggests the world emissions need to start falling by 2015 if we are to avoid "dangerous" climate change. Under this book's proposed policy, in 2015 Australia would have approximately 15 of these facilities in operation (about 2 per state) and around another 15 under construction. We would have around 170 by 2040, and be operating on 100% renewable energy. The USA would have about 3 per state operational by 2015.

(60)

(61)

(62)

The next world climate change accord

If the science is correct, climate change is the most serious danger that modern humanity has faced. This book deliberately does not describe the extent of the risks in much detail.

This is a time when the world needs great leaders. Fifty years ago President Kennedy announced that the USA would land a man on the moon within 10 years. We need that kind of leadership now.

There was no point landing a man on the moon 50 years ago, if we are going to greatly harm our own planet 50 years from now, largely through complacency or inadequate risk management. It was scientists that put men on the moon, not politicians, and not the general public. We need to listen to our scientists.

Both the Australian and USA Governments have been recently more focused on the issue of heath reform, which is important, but not as important as climate change is forecast to be. As the USA gets past the efforts involved in its new health reform legislation, as well as new initiatives in nuclear security and immigration, the US Congress might next focus on the big issue of climate change.

In line with the policies that are suggested in this book our newly announced goal should be for 100% renewable energy and being carbon neutral or better by 2040.

This is a significant step up from modest emission reductions targets that we are currently setting and are not thought to give us a 50% chance of avoiding "dangerous" levels of climate change. We want a better than 50% chance.

The world's current largest emitting nations are the USA, China, India, Russia and Japan. Australia is one of the world's highest emitting nations per capita, just ahead of the USA.

The next World climate change accord should flow on from the Copenhagen discussions and should start with the six leaders of these five largest emitting nations plus Australia. This would address over 50% of the world's emissions. It could extend to other countries on an opt-in basis. As and when the risks become more obvious, the world's public will demand it.

Meetings for the next World Accord that includes the world's five largest emitters, could be hosted here in Australia.

This book summarises the science of climate change and examines a variety of climate change policies. It suggests and costs a large-scale direct-action plan. It contains only a little discussion of the risks, but explains that we need to urgently 'stomp-out the fuse". Whether or not a price is put on carbon, this book explains why this will only be achieved in the required timeframes with a very large-scale direct-action plan.

The large scale direct-action plan outlined in this book is necessarily high level at this stage of planning. Attachments to this book contain calculations of cost of 100% renewable energy for the USA, China and the world as a whole by 2040. A draft plan for the USA to become carbon neutral by 2040 is also provided. These plans will need more detailed development and will necessarily evolve over time. Also, details will vary between nations, based on land area, land usage, climate and population.

This is the decade when we need to start aggressively reducing carbon emissions. We have the ability to do so. Will we be the responsible generation, or not?

Based on an awareness of the risks, and of the most likely truth, how long should it take us to decide to embark on such a programme of action? About 1 second. 5 or 10 at the most.

References

1. Ian Plimer, Heaven + Earth – Global Warming: The Missing Science, Connorcourt Publishing, 2009.
2. NASA, http://en.wikipedia.org/wiki/File:Nasa-giss_1880-2009_global_temperature.svg
3. http://en.wikipedia.org/wiki/File:Mauna_Loa_Carbon_Dioxide-en.svg, This file is licensed under the Creative Commons Attribution-Share Alike 3.0 Unported license
4. Wikipedia - Greenhouse effect.
5. "Water vapour: feedback or forcing?". RealClimate. 6 April 2005. http://www.realclimate.org/index.php?p=142.
6. Kiehl, J. T.; Kevin E. Trenberth (February 1997). "Earth's Annual Global Mean Energy Budget" (PDF). *Bulletin of the American Meteorological Society* **78** (2): 197–208.
7. Robert A Rohde, Dragons Flight, Global Warming Art, www.globalwarmingart.com, http://en.wikipedia.org/wiki/File:Climate_Change_Attribution.png
8. Sunshine speeded 1940s Swiss glacier melt, http://www.physorg.com/news180024364.html
9. Robert A Rohde, Dragons Flight, Global Warming Art, www.globalwarmingart.com, http://www.globalwarmingart.com/wiki/File:2000_Year_Temperature_Comparison_png
10. US Department of Commerce, National Oceanic and Atmospheric Administration, http://oceanservice.noaa.gov/education/pd/climate/factsheets/areincrease.pdf
11. NASA, http://data.giss.nasa.gov/gistemp/graphs/Fig.C.lrg.gif
12. United States geological Survey http://www.globalwarmingart.com/wiki/File:Carbon_Derived_Solar_Change_png
13. http://www.skepticalscience.com/acrim-pmod-sun-getting-hotter.htm
14. www.geocraft.com, http://www.geocraft.com/WVFossils/Carboniferous_climate.html
15. http://www.skepticalscience.com/co2-lags-temperature.htm, Vostok ice core records for carbon dioxide concentration (*Petit 2000*) and temperature change (*Barnola 2003*). Carbon dioxide Information Analysis Centre, http://cdiac.ornl.gov
16. Central England Temperature Series, Lord Monckton, Presentation - Apocalypse ? NO!, 2009.
17. Leland Mcinnes, Wikipedia - Dansgaard-Oeschger event, http://en.wikipedia.org/wiki/File:Grip-ngrip-do18-closeup.png. *This file is licensed under the Creative Commons Attribution ShareAlike 3.0 License*
18. Robert A Rohde, Dragons Flight, Global Warming Art, www.globalwarmingart.com, http://commons.wikimedia.org/wiki/File:Holocene_Temperature_Variations.png
19. Schiermeier, Q., No solar hiding place for greenhouse skeptics. Nature, 2007. 448: p. 8-9.
20. Dr Roy Spencer, Clouds Dominate CO2 as a climate driver since 2000, http://wattsupwiththat.com/2010/01/13/spencer-clouds-dominate-co2-as-a-climate-driver-since-2000/
21. http://climate4you.com/GlobalTemperatures.htm
22. El Nino – Global Temperature Driver, Global Warming Science, http://www.appinsys.com/GlobalWarming/GlobalElNino.htm
23. Rosentod, Marsupilami, United States Geological Survey, AMO Oscillation Timeseries with a 12 month moving average 1956-2009, http://en.wikipedia.org/wiki/File:Amo_timeseries_1856-present.svg
24. US National Oceanic and Atmospheric Administration, Monthly Values for the PDO index: 1900- September 2009, http://en.wikipedia.org/wiki/File:Pdoindex_1900_present.png
25. Wikipedia - Snow ball Earth.
26. Rasool, I.; De Bergh, C. (Jun 1970). "The Runaway Greenhouse and the Accumulation of CO2 in the Venus Atmosphere". *Nature* **226** (5250):
27. Lord Monckton, Presentation - Apocalypse? NO!, 2009.

28. Climate Shifts, http://www.climateshifts.org/?p=4369, ABC TV Mediawatch, J.M. Lough, AIMS, February 2010.
29. Carl Sagan, "Pale Blue Dot", Headline Book Publishing, London, 195
30. Stewart Needham., The potential for renewable energy to provide baseload power in Australia, Research paper no 9. 1008-09, Parliament of Australia, Parliamentary Library.
31. IPCC, 2007: Summary for Policymakers. In: Climate Change 2007: Impacts, Adaptation and Vulnerability. Contribution of Working Group II to the Fourth Assessment Report of the Intergovernmental Panel on Climate Change, M.L. Parry, O.F. Canziani, J.P. Palutikof, P.J. van der Linden and C.E. Hanson, Eds., Cambridge University Press, Cambridge, UK, 7-22.
32. www.petitionproject.com
33. Sahara Desert Greening Due to Climate Change ?, National Geographic, http://news.nationalgeographic.com/news/2009/07/090731-green-sahara.html
34. IPCC, 2007: Summary for Policymakers. In: Climate Change 2007: Mitigation. Contribution of Working Group III to the Fourth Assessment Report of the Intergovernmental Panel on Climate Change [B. Metz, O.R. Davidson, P.R. Bosch, R. Dave, L.A. Meyer (eds)], Cambridge University Press, Cambridge, United Kingdom and New York, NY, USA., http://www1.ipcc.ch/pdf/assessment-report/ar4/wg3/ar4-wg3-spm.pdf
35. Mini-ice age could hit in a manner of months, not years, http://www.gizmag.com/mini-ice-age-hit-in-months/13489/
36. Venus image - Image processing by R. Nunes, http://www.astrosurf.com/nunes
37. R.E.H. Sims, R.N. Schock, A. Adegbululgbe, J. Fenhann, I. Konstantinaviciute, W. Moomaw, H.B. Nimir, B. Schlamadinger, J. Torres-Martínez, C. Turner, Y. Uchiyama, S.J.V. Vuori, N. Wamukonya, X. Zhang, 2007: Energy supply. In Climate Change 2007: Mitigation. Contribution of Working Group III to the Fourth Assessment Report of the Intergovernmental Panel on Climate Change [B. Metz, O.R. Davidson, P.R. Bosch, R. Dave, L.A. Meyer (eds)], Cambridge University Press, Cambridge, United Kingdom and New York, NY, USA.
38. Australian Treasury, Australia's Low Pollution Future: The Economics of Climate Change Mitigation, http://www.treasury.gov.au/lowpollutionfuture
39. Graeme Pearman Consulting Pty Ltd, Climate Change Risks in Australia under Alternative Emissions Futures, report for the Australian Government Department of Treasury
40. Labor's 2020 target for a renewable energy future, Election 2007 policy document, October 2007
41. Coalition's Direct Action Plan, http://www.nationals.org.au/DirectActionPlan.aspx
42. Reducing Australia's Emissions, Department of Climate Change and Energy Efficiency, web site http://www.climatechange.gov.au/en/government/reduce.aspx
43. McLennan Magasanik Associates, Report to Federal Treasury Impacts of the Carbon Pollution Reduction Scheme on Australia's Electricity Markets, 2008, www.treasury.gov.au
44. Frenc M. Miskolczi, Greenhouse effect in semi-transparent planetary atmospheres, Quarterly Journal of the Hungarian Meteorological Service, Vol111, No 1, January-March 2007
45. Larry Lohmann, Carbon trading – A critical conversation on climate change, privatization and power, The Corner House, development dialogue no 48. September 2006, www.thecornerhouse.org.uk
46. Tamra Gilbertson and Oscar Reye, Carbon Trading – How it works and why it fails, Dag Hammarskjold Foundation, Uppsala, , Critical Currents Occasional Paper Series No7, November 2009
47. Australian Government, Carbon Pollution Reduction Scheme, Australia's Low Pollution Future – White paper Volume 1, December 2008
48. Department of Climate Change, Australia's Fifth National Communication on Climate Change, 2010
49. Australian Greens web site, http://greens.org.au/

50. Four Peaks, Solar annual 2008, http://www.photonconsulting.com/files/sa_2008_exec_summary.pdf
51. Gerson, Lehrman Group, http://www.tnsolarenergy.org/downloads/PV%20Cost%20Reductions%20bright%20outlook.pdf
52. The Australian Green's Safe Climate Bill, http://greensmps.org.au/the-safe-climate-bills
53. BIOchar, http://sites.google.com/site/yarravalleyclimateactiongroup/forest-biomass-derived-biochar-can-profitably-reduce-global-warming-and-bushfire-risk
54. Berney, Banks and Mackay, BIO-Agriculture – A solution to Global Climate Change, http://bio-agriculture.org/home.html
55. Australian Government, Australian Natural Resources, Atlas, http://www.anra.gov.au/topics/land/landuse/index.html
56. Guardian, 'Biochar' goes industrial with giant microwaves to lock carbon in charcoal, http://www.guardian.co.uk/environment/2009/mar/13/charcoal-carbon
57. George Monbiot, Woodchips with everything, http://www.monbiot.com/archives/2009/03/24/woodchips-with-everything/
58. Institute of Public Affairs, So you have bought an emissions trading scheme, http://www.ipa.org.au/publications/1359/so-now-you-have-bought-an-emissions-trading-scheme
59. Nuclearinfo.net, Everything you want to know about nuclear power, http://nuclearinfo.net/Nuclearpower/WebHomeCostOfNuclearPower
60. http://www.fourgreensteps.com/infozone/world-news/general/cobra-energy-plans-massive-1-billion-solar-thermal-plant-for-australia
61. http://www.brightsourceenergy.com/images/uploads/Brightsource_Ivanpah_Fast_Facts.pdf
62. www.esolar.com
63. http://www.coloradotrees.org/benefits.htm#pollutants. A total of 300 trees can counter balance the amount of pollution one person produces in a lifetime.
64. Friends of the Earth, Subprime Carbon – Rethinking the world's largest new derivatives market, March 2009, www.foe.org
65. Stephen Schneider, Armin Rosencranz, Michael Mastrandrea, Kriston Kuntz-Duriseto, Climate Change Science and Policy, island Press, 2010
66. Ross Garnaut, The Garnaut Climate Change Review Final report, Cambridge University Press, 2008
67. Helwig-Larsen and Bull, ZeroCarbonBritain – an alternative energy strategy, Centre for Alternative Technology, 2007, www.cat.org.uk
68. http://science.howstuffworks.com/lake-nyos.htm/printable
69. The Carbonsafe Alliance submission to the inquiry into Geosequestration Technology, http://www.aph.gov.au/house/committee/scin/geosequestration/subs/sub38.pdf
70. Virgin Earth Challenge, http://www.virgin.com/subsites/virginearth
71. Climate Works, Low Carbon Growth Plan, http://climateworksaustralia.com
72. Stanford Law School, http://www.law.stanford.edu/news/details/1722/Stanford%20Study%20May%20Stir%20Debate%20On%20Limiting%20Costs%20In%20Climate%20Bill/
73. Barbara Haya, UC Berkeley, http://www.co2offsetresearch.org/PDF/AdditionalityLackCDM.pdf

Suggested further reading, mainly from the web

The **IPCC's Fourth Assessment Report** (31, 34) is available on the web and is definitely worth some look. It is criticised for being too large.

The Australian **Garnaut** report (66) is also a very large and comprehensive report on the risks and options for Australia. It has very good analysis of options for land-use, transportation and energy. It extensively advocates emissions trading, but also explains circumstances in which emissions trading should not be used. It is about 700 pages.

Climate Change Science and Policy (65) is available from CSIRO publishing for around $100. It is a 500 page American book written by many climate change experts and covers 49 different topics, including analyses of what is going on in a range of countries, such as the USA, China and Australia.

Australia's Low Pollution Future (38) is a description of the proposed emissions trading approach for Australia. There is a summary document and a main document.

Climateworks, **Low carbon Growth Plan** (71) identifies many emissions reductions opportunities in Australia but sees a low potential for renewable energy, particularly solar, in the next 10 years.

ZeroCarbonBritain is a similar book to this one, but focusing on Britain, and is available as a free download on the web. It is 100 pages. Rather than a book written by one lay-person, as is this one, it was produced by about 60 experts and nearly as many reviewers. It advocates a target of 100% renewable energy and 100% emission reduction for Britain in 20 years, rather than the 30 years suggested in this book for Australia. It does not attempt to cost the proposal, as this book does. It has very interesting descriptions of options for energy, transportation and agriculture – the problems are very similar to Australia's but some of the proposed solutions are a bit different. Some of the interesting ideas are that electric cars will each store enough power for an average home for around two days, or as a national fleet, for a similar period for the country. It therefore suggests cars as a good means of storing large amounts of energy, as well as pumping water up-hill for hydro-electric power, as we do here in Australia. It suggests a different method of carbon trading where individuals hold "Tradeable Energy Quotas", and also an equitable approach for the world of "Convergence and Contraction", where all countries aim for a common future per capita carbon intensity.

Carl Sagan's **Pale Blue Dot** (29) describes our potential future in space, and discusses the practicalities and ethics involved in geo-engineering other planets to make them habitable.

James Hanson, one of the world's leading climate scientists, has written a book about climate change. It is more detailed than this one and also outline the risks in more detail. It is available on Amazon.

The "*Pale Blue Dot*". From Wikipedia : Seen from 6 billion kilometres (3.7 billion miles), Earth appears as a tiny dot (the blueish-white speck approximately halfway down the brown band to the right). The **Pale Blue Dot** is a photograph of planet Earth taken in 1990 by Voyager 1 from a record distance, showing it against the vastness of space. By request of Carl Sagan, NASA commanded the Voyager 1 spacecraft, having completed its primary mission and now leaving the solar system, to turn its camera around and to take a photograph of Earth. It was subsequently used by Sagan as the title of his 1994 book of the same name. In 2001, it was selected by Space.com as one of the top ten space science photos.

Attachments

President Kennedy's speech to Congress on landing a man on the moon

"IF we are to win the battle that is now going on around the world between freedom and tyranny, the dramatic achievements in space which occurred in recent weeks should have made clear to us all, as did the Sputnik in 1957, the impact of this adventure on the minds of men everywhere, who are attempting to make a determination of which road they should take. Since early in my term, our efforts in space have been under review. With the advice of the Vice President, who is Chairman of the National Space Council, we have examined where we are strong and where we are not. Now it is time to take longer strides--time for a great new American enterprise--time for this nation to take a clearly leading role in space achievement, which in many ways may hold the key to our future on Earth.

I believe we possess all the resources and talents necessary. But the facts of the matter are that we have never made the national decisions or marshaled the national resources required for such leadership. We have never specified long-range goals on an urgent time schedule, or managed our resources and our time so as to insure their fulfillment.

Recognizing the head start obtained by the Soviets with their large rocket engines, which gives them many months of lead-time, and recognizing the likelihood that they will exploit this lead for some time to come in still more impressive successes, we nevertheless are required to make new efforts on our own. For while we cannot guarantee that we shall one day be first, we can guarantee that any failure to make this effort will be our last. We take an additional risk by making it in full view of the world, but as shown by the feat of astronaut Shepherd, this very risk enhances our stature when we are successful. But this is not merely a race. Space is open to us now; and our eagerness to share its meaning is not governed by the efforts of others. We go into space because whatever mankind must undertake, free men must fully share.

I therefore ask the Congress, above and beyond the increases I have earlier requested for space activities, to provide the funds which are needed to meet the following national goals:

First, I believe that this nation should commit itself to achieving the goal, before this decade is out, of landing a man on the Moon and returning him safely to the Earth. No single space project in this period will be more impressive to mankind, or more important for the long-range exploration of space; and none will be so difficult or expensive to accomplish. We propose to accelerate the development of the appropriate lunar space craft. We propose to develop alternate liquid and solid fuel boosters, much larger than any now being developed, until certain which is superior. We propose additional funds for other engine development and for unmanned explorations--explorations which are particularly important for one purpose which this nation will never overlook: the survival of the man who first makes this daring flight. But in a very real sense, it will not be one man going to the Moon--if we make this judgment affirmatively, it will be an entire nation. For all of us must work to put him there.

Secondly, an additional 23 million dollars, together with 7 million dollars already available, will accelerate development of the Rover nuclear rocket. This gives promise of some day providing a means for even more exciting and ambitious exploration of space, perhaps beyond the Moon, perhaps to the very end of the solar system itself.

Third, an additional 50 million dollars will make the most of our present leadership, by accelerating the use of space satellites for world-wide communications.

Fourth, an additional 75 million dollars--of which 53 million dollars is for the Weather Bureau--will help give us at the earliest possible time a satellite system for world-wide weather observation.

Let it be clear--and this is a judgment which the Members of the Congress must finally make--let it be clear that I am asking the Congress and the country to accept a firm commitment to a new course of action--a course which will last for many years and carry very heavy costs: 531 million dollars in fiscal '62--an estimated seven to nine billion dollars additional over the next five years. If we are to go only half way, of reduce our sights in the face of difficulty, in my judgment it would be better not to go at all.

Now this is a choice which this country must make, and I am confident that under the leadership of the Space Committees of the Congress, and the Appropriating Committees, that you will consider the matter carefully.

It is a most important decision that we must make as a nation. But all of you have lived through the last four years and have seen the significance of space and the adventures in space, and no one can predict with certainty what the ultimate meaning will be of mastery of space.

I believe we should go to the Moon. But I think every citizen of this country as well as the Members of the Congress should consider the matter carefully in making their judgment, to which we have given attention over many weeks and months, because it is a heavy burden, and there is no sense in agreeing or desiring that the United States take an affirmative position in outer space, unless we are prepared to do the work and bear the burdens to make it successful. If we are not, we should decide today and this year.

This decision demands a major national commitment of scientific and technical manpower, material and facilities, and the possibility of their diversion from other important activities where they are already thinly spread. It means a degree of dedication, organization and discipline which have not always characterized our research and development efforts. It means we cannot afford undue work stoppages, inflated costs of material or talent, wasteful interagency rivalries, or a high turnover of key personnel.

New objectives and new money cannot solve these problems. They could in fact, aggravate them further--unless every scientist, every engineer, every serviceman, every technician, contractor, and civil servant gives his personal pledge that this nation will move forward, with the full speed of freedom, in the exciting adventure of space."

0 0 0 0 0 0 0 0 0

Footnote: In 1961 the GDP of the USA was US$531B. The commitment of $9B over the next five years was 0.3% of GDP per annum. Ultimately the total Apollo programme cost of $24B over 10 years to 1972 was 0.6% of the USA's 1961 GDP.

The calculated cost of replacing the USA's electricity generation with renewable energy is a much larger exercise than the Apollo programme. At $US4 Trillion, it is however very similar to some estimates of the cost of the Iraq war at US$3 Trillion, but is cheaper in that this expenditure would occur over 30 years.

Interestingly, NASA now assists greatly in furthering climate science.

How much would it cost the USA to implement 100% renewable energy?

It is cheaper per capita but slightly higher as a proportion of GDP in the USA than it is in Australia. The USA's current energy usage is about 16 times Australia's at approximately 4000 TWh, or 4,000,000 GWh.

100% renewable energy based on current usage requires 4,000,000 Gwh.

Using solar only, this would require 2,666 1GW facilities (4,000,000 GWh/1500 GWh = 2,666).

At $1.5B each this is $4,000B.

The USA's population is approximately 400M.

Over 30 years, the above cost is **only $333 per person per year** ($4,000,000M/400M/30 years).

The USA's gross domestic product is about $14,000B.

Over 30 years, the above cost is **only 0.9% of GDP p.a.** ($4,000B/30 years/$14,000B).

The above calculation uses today's or near future prices, which will reduce, but has not factored in energy usage growth. In the USA, drop in prices will probably exceed usage growth, making the above an over-estimate. The cost per person above is an average. Depending on how it is funded, those that consume more would pay more, and those that consume less would pay less.

This book has concentrated on the Australian political environment. A USA edition of this book could be developed that focuses on the USA political environment.

How much would it cost China to implement 100% renewable energy?

China's energy usage is growing rapidly. In 20 years it will be 7,500 TWh, or 7,500,000 GWh which is about 30 times Australia's.

100% renewable energy based on current usage requires 7,500,000 Gwh.

Using solar only, this would require 5,000 1GW facilities (7,500,000 GWh/1500 GWh = 5000).

At $1.5B each this is $7,500B.

China's population is approximately 1,500M.

Over 30 years, the above cost is $166 per person per year ($7,500,000M/1,500M/30 years).

China's gross domestic product is about $4,300B.

Over 30 years, the above cost is 6% of GDP p.a. ($7,500B/30 years/$4,300B).

The above calculation uses today's or near future prices which will reduce. It might be cheaper in China to implement solar facilities and so the above might be an over-estimate. China's GDP has doubled since 2004/2005, and so this investment is a small proportion of China's overall economic growth.

How much would it cost the World to implement 100% renewable energy?

It is cheaper per capita but much higher as a proportion of GDP for the world than it is in Australia. The world's energy usage is 16,800 TWh, or 16,800,000 GWh (65 times Australia).

100% renewable energy based on current usage requires 16,800,000 Gwh.

Using solar only, this would require 11,200 1GW facilities (16,800,000 GWh/1500 GWh = 11,200).

At $1.5B each this is $16,800B.

The world's population is approximately 7,000M.

Over 30 years, the above cost is $80 per person per year ($16,800,000M/7,000M/30 years).

The world's gross domestic product is about US$60,000B.

Over 30 years, the above cost is 0.9% of the world's GDP p.a. ($16,800B/30 years/$60,000B).

The above calculation uses today's or near future prices, which will reduce, has not factored in any energy usage growth. Drop in prices will probably exceed usage growth, making the above an over-estimate. The cost per person above is an average. Depending on how it is funded, those that consume more would pay more, and those that consume less would pay less.

Scaling the proposed Australian direct-action plan to the USA

The following is a draft Australian 30 year direct-action expenditure plan scaled to the larger size of the USA. The USA emits 5,752 Mt of CO2 p.a. (2006). If 100% of this were taxed at say $30/tonne, the revenue raised would be $172B p.a.

To raise the same revenue through the USA's electricity generation (4000 TWh p.a.) requires:

$$\frac{\$172B}{4000\ TWh} = \frac{\$1}{23\ KWh} = \textbf{4.3 cents per kilowatt hour}$$

At $172B p.a. the total 30 year expenditure is a little over $5,000B in total in today's dollars.

$4000B	**Complete replacement of electricity supply with renewable energy**. This major capital project creates hundreds of thousands of jobs. Gas fired power stations may be maintained as backup power.
$50B	**Reforestation of the USA.** At $10 per tree this is 5 billion trees or 12 trees for every American. These trees will create a carbon sink over their growing lifetime and will contribute greatly to America being net carbon neutral. This approach is very effective. Reference 62 suggests it takes 300 trees to cancel out a person's lifetime emissions. More analysis of the potential of this approach is required. Under this policy a saving of 250 MT per annum over existing forests will be targeted .
$50B	**Soil abatement trial.** This approach will be trialed with the aim of achieving a target of 1000 MT of abatement pa. Conservation tilling and BIOchar approaches will be among the approaches trialed.
$80B	**CO2 capture research.** This substantial budget will be used to fund research into clean approaches to reducing CO2 in the atmosphere, with the aim of developing a commercially viable approach for world-wide application. The budget for this activity anticipates and includes the building of the first large scale commercial facilities.
$100B	Additional agricultural assistance, in a form to be determined, to help reduce emissions in that area.
$70B	Assist development of **hydrogen and electric green cars**.
$100B	Assist development of **hydrogen refueling facilities** for green cars.
$750B	**Green car subsidy.** Purchase of 50 million new green cars built in the USA will be subsidised for new car buyers. The average subsidy over 30 years will be $15,000 each, but starting at $30,000 each in year 1 and reducing over 30 years. Initially a $45,000 electric car will be available for $15,000, and a more advanced and powerful $75,000 car will be available for $45,000, and a $120,000 hydrogen luxury car will be available for $90,000. $75B will subsidise 5 million cars. There will be a waiting list for these cars. 50,000 of the cheaper cars will be used as public cars in major towns and cities. After the subsidies cease in 30 years only green cars will be available for sale in the USA. Along with other improvements to transport efficiency that will occur over this time, a saving of 500MT per annum is targeted.
$100B	**The icing on the cake.** As part of the new world Accord, the USA, with contributions from Australia, China and other participating countries will **buy out deforestation activities in the Amazon** and other major world rainforests, thus saving an estimated 20% of the worlds net carbon emissions. Taking credit for 2/3 of this action puts the USA at around 500% net carbon negative. Suggestions for this very effective initiative are at reference 53 and urgently require more investigation.
$5300B	**Total funds required over 30 years**

The 4.3 cents per kilowatt hour goes a surprising very long way, if spent effectively.

The draft plan for the USA targets the same types of emission reduction activities as in the Australian draft plan presented in this book. In many areas the proposed expenditure and benefits are the same as proposed for Australia but multiplied 10. An exception is forestation. The USA's land area is similar to Australia's. Nevertheless the initially suggested budget and targeted benefits is 5 times Australia's. The USA has less desert than Australia.

In summary, the 30 year plan includes implementing 100% renewable energy, planting a lot of trees and accelerating the adoption of electric and hydrogen cars. It trials soil abatement and other forms of agricultural carbon abatement. In a combined action with other nations, it stops deforestation in the Amazon and other countries.

It accelerates works alongside other initiatives that will occur over the same time period, such as increased building efficiencies, hydrogen passenger aircraft, and improvements in waste management.

This kind of direct action approach can co-exist alongside emissions trading, if the pool of permits funding were directed towards such scale of direct action.

The emissions saving targets under this proposed policy are thus as follows, being proportions of the USA's current emissions.

Electricity	over 2500 MT pa
Soil abatement	at least 1000 MT pa from this and other improved agricultural practices including "carbon farming"
Forestry	at least 250 Mt pa, from accelerated onshore reforestation
Transport	growing to 500 MT pa, based on expected advancements and with acceleration from proposed electric car replacement subsidies
Other efficiencies	500 Mt pa. These will be achieved across a range of sectors (building efficiencies, waste management, mining)

This totals 4750 Mt per annum, which is close to carbon neutral onshore based on the USA's current emissions of around 5752 Mt p.a. It is possible that with more thorough analysis of improved agriculture, soil and forestry based sequestation approaches, a better result than this could be achieved.

Adding on benefits of the 2/3 contribution to stopping rain forest reforestation off-shore in a range of countries, which at around 20% of the world's emissions is similar to the USA's emissions, this takes the USA to about 66% carbon negative.

Garnaut report tables on agriculture and land use abatement potential

Reproduced with the standard permissions from the Garnaut Climate Change Review- Copyright of the Commonwealth of Australia 2008.

Table 1. Technical potential for CO2 removal by soil—selected estimates

Activity	Location	Carbon dioxide removal estimate[a]	Source
Conservation and sustainable land management practices	Worldwide	3.3 Gt (± 1.1 Gt) CO_2 per year (for 50 years)	Lal (2004)
		1.6–3.2 Gt CO_2 per year (for 50 years)	Paustian et al. (2004)
Adoption of sustainable stocking rates, rotational grazing and seasonal use grazing practices	United States rangelands (previously degraded)	0.3–1.3 t CO_2 per ha per year	Chicago Climate Exchange (2008)
Zero till or minimum tillage	United States	0.5–1.5 t CO_2 per ha per year	Chicago Climate Exchange (2008)
Conservation tillage	Australia-wide	25% increase in carbon retained, compared to conventional tillage	Valzano et al. (2005)
Conservation tillage	South-eastern Australia	43.3-46.6 t CO_2 per ha over 20 years in high rainfall region (on average between 2.1 and 2.3 t per year)	Grace et al. (2004)
Changes in cropping practice	New South Wales	2 t CO_2 per ha per year	Chan (unpublished)
Sowing of crops into perennial pastures (growth of perennial grasses alongside crops)	Trial sites in New South Wales with good soils	5-10 t CO_2 per ha per year (up to 20–30 t under good conditions)	C. Jones (2008, pers. comm.)
Pasture cropping (pasture type: Rhodes, Lucerne, Siratro, Bambatsi)	Loam soil, 'northern agriculture region', Western Australia	5.2 t CO_2 per ha per year	T. Wiley (2008, pers. comm.)

a Soil carbon will eventually reach a new equilibrium, and carbon removals in soil will cease. Removals may continue for 20–50 years before a new equilibrium is reached.

Table 2. Potential for emissions per annum reduction and/or removal from Australia's agriculture, forestry and other land use sectors

Process	Potential	Key assumptions
Land clearing (deforestation)	Emissions reduction potential of 63 Mt CO_2-e per year on an ongoing basis	Land clearing ceases (resulting in zero emissions from deforestation).
Enteric emissions from livestock	Emissions reduction estimated at 16 Mt CO_2-e per year on an ongoing basis	Based on either deployment of anti-methanogen technology for ruminant livestock, or shifting of meat production from a minority proportion (7 million cattle and 36 million sheep) of ruminant livestock by kangaroos.
Removal by soil—cropped land	Removal potential of 68 Mt CO_2-e per year for 20–50 years	Conservative management changes assumed (e.g. conservation tillage), not pasture cropping. Changed practices implemented on all cropped land (38 million ha).
Removal by soil—high-volume grazing land	Removal potential of 286 Mt CO_2-e per year for 20–50 years	Based on the Chicago Climate Exchange for changed practices to rehabilitate previously degraded rangelands. Changed practices implemented on all grazing land (358 million ha).
Restoration of mulga country	Up to 250 Mt CO_2-e per year for several decades	Comprehensive restoration of degraded, low-value grazing country in arid Australia.
Nitrous oxide emissions from soil	Reduction potential of 0.3 Mt CO_2-e per year for 20–50 years	Improved fertiliser management practised on all agricultural soils.
Reduction in emissions from savanna burning	Reduction of 5 Mt CO_2-e per year on an ongoing basis	Assumes annual emissions from savanna fires are 10 Mt (average from period 1990–2006). Complete reduction of savanna fire is not desirable or feasible; a 50% reduction is assumed, through management.
Removal by post-1990 forests	Emissions removal potential of 50 Mt CO_2-e by 2020	Assumes Australia will have around 2 million ha of Kyoto-compliant (post-1990) plantations (including wood production plantations and specific carbon plantations) by 2020.
Removal by pre-1990 eucalypt forests	Emissions removal potential equivalent to 136 Mt CO_2-e per year (on average) for 100 years	Current carbon stocks in logged forests are about 40% below carrying capacity. Timber harvesting and other human disturbances cease in study area (14.5 million ha). Landscape growth potential has not been degraded by land use activities.
Carbon farming (plantations)	Emissions removal potential of 143 Mt CO_2-e per year for 20 years	Using 9.1 million ha of land where returns would be more than $100 per ha per year better than current land use, with water interception less than 150 mm per year and permit price of $20 per tonne CO_2-e.
Biofuel production	Up to 44 Mt CO_2-e per year on an ongoing basis	Replacement of all fossil fuel diesel with biodiesel. More than 550 000 ha required for production (cultivating algae as a feedstock) or more than 10 million ha (using other plants).

Sources: Beeton (pers. comm.); Chicago Climate Exchange (2008); DCC (2008a); de Klein & Eckhard (2008); Grace et al. (2004); IPCC (2006); Mackey et al. (2008); NAFI & TPA (2007); Chan (unpublished); Polglase et al. (2008); Russell-Smith et al. (2004); Wilson & Edwards (2008).

Disclaimers

Climate change is an enormous and overwhelming topic. Whilst the author has endeavored to ensure reasonable accuracy and completeness of all statements in this book, this cannot be guaranteed. All statements in this book should therefore be regarded as the opinion of the author, and readers are advised to examine the policy documents and other papers referenced in this book to gain a complete and accurate picture of all ideas, policies and scientific statements that are summarised in this book.

While reasonable efforts have been made to ensure that the contents of this publication are factually correct, the author does not accept responsibility for the accuracy or the completeness of the contents, and shall not be liable for any loss or damage that may be occasioned directly or indirectly through the use of, or reliance on the contents of this publication.

Possible Inaccuracies

The following measures are easily confused and may have been confused in parts of this book.

A$1 is approximately equal to US$1. In the capital costs of 100% renewable energy issues of Australian exchange rate have been ignored for the purposes of the order-of-magnitude calculations.

One tonne of carbon = around 3.5 tonnes of CO2. Most figures in this book about emission levels and trading permits relate to tonnes of CO2. In some circumstances measures relate to tonnes of carbon. One tonne of carbon saves around 3.5 tonnes of CO2. The author is not aware that any such confusion has been made in this book.

CO2-e. Discussion of CO2 levels in the atmosphere are primarily about CO2 itself. e.g. CO2 is currently around 390 ppm in the atmosphere. Most target figures in planned emissions trajectories such as 450 ppm and 550 ppm are about CO2-e (CO2 equivalent) which is a measure of the effect of all additional greenhouse gasses, including methane, in terms of how much warming they produce collectively in relation to CO2. The author's understanding is that we are currently somewhere around 420ppm of CO2-e, thus not very far from the 450 ppm stablilisation limit that is thought to give us about a 50% chance of keeping warming less than 2 degrees. Stabilisation at 450 ppm means that we will overshoot that limit but hopefully get it back.

Warming probabilities. Different literature provides different probabilities of likelihood of warming at different CO2-e levels. The 50% chance of 450 ppm leading to 2 degrees of warming is an average estimation. Some literature is more pessimistic than this.

Emission target baseline years. Various reduction targets are from different year baselines, such as from 1990 levels, 2000 level or 2005 levels for example. The particular year baseline is generally stated in this book but sometimes not for brevity.

Kyoto protocol. The author is not an expert of Kyoto rules in areas such as forestation and de-forestation. These rules appear complex. They might not be represented in this book with absolute accuracy.

USA policy situation. The author is not fully up to speed with latest USA policy thinking, and might not have portrayed the USA political situation accurately.

Updates and Feedback

Updates and further information about this book can be found at the following web site.

www.climate-change-explored-explained.com

The author can be contacted at c-c-e-e@grapevine.com.au.

18 May 2011 Fourth letter

The Hon Julia Gillard MP
Prime Minister of Australia
PO Box 6022
Parliament House
Canberra ACT 2600

The Potential Climate Policy Impasse

How Current and Future Generations Should Pay for Renewable Energy

Dear Ms Gillard

The Potential Climate Policy Impasse

Australia currently faces a potential climate policy impasse. A battle of different ideas, ideologies and policies ensues. A lack of public and political party consensus on the best approach means we are unlikely to agree and commence a good program of action. Our Government is unstable as a result.

The **Greens** advocate aggressive action, including a carbon price, a carbon floor price, a plan for 100% renewable energy, and feed-in tariffs to subsidise the current high cost of large-scale renewable energy. The Greens also advocate more aggressive targets for emissions reductions, and claim that carbon pricing by itself does not provide an adequate approach.

The **Coalition** advocates a cheap approach to minimal action. The Coalition proposes to spend only $12B over 10 years. Their policy suggests cheap approaches to match Labor's minimum proposed 5% emission reduction targets, and avoid any carbon pricing.

Labor sits in the middle. A low starting carbon price will be proposed, with the belief that over time a market-based mechanism will efficiently bring about an adequate level of action. The carbon price will rise over time. The carbon price is unpopular with voters. Real action will be slow.

At the heart of the debate lie differences about the effectiveness of carbon pricing, and also the level of urgency and risk management required to protect us from future major environmental impacts, and also possible major economic impacts of peak oil.

For the public at large, there is an unwillingness to pay more now for a substantial level of action to protect against future risks. A lack of evidence for the need, and the desire of humans to look after their own backyards, will continue to prevent an appropriate level of risk management.

The Greens want to start paying for substantial action now. A carbon price, a solution currently proposed, is based on the principle of starting to pay a bit now, and paying more later. However the starting prices under discussion are not sufficient to fund renewable energy and may not be sufficient to be wasted on gas. Whether a carbon price is effective is debatable.

The heart of the question is how much current generations should pay now in order to best protect future ones.

My letter of 19 April 2011, as well as other analysis, shows that a rapid and aggressive implementation of renewable energy, over 10 to 30 years, will be much cheaper than staying with coal, with savings to Australia of between one to two trillion dollars over that timeframe.

However, the current generation is unwilling to pay the cost of starting. As a result we will not properly start the process, to save money over a period of decades, and provide protection against future environmental and economic risks.

A Climate Policy Solution – How Current and Future Generations Should Pay

A mechanism is required for both current generations and future generations to share the costs of ensuring our environmental and economic survival. The benefits can easily be made clear. A way to pay for it needs to be agreed.

The Greens advocate feed-in tariffs for large-scale renewable energy. These make the public pay the high capital cost of renewable energy now. The Coalition points out their impact on electricity prices. Feed-in tariffs are intended to give renewable energy generators a viable pay-back period for investments made now.

It is expected that feed-in tariffs will reduce over time, as renewable energy costs reduce, but they require the public to start paying the high price now.

As the feed-in tariffs reduce, fossil-fuel based prices will rise, to the point where they become equal, and the tariff is now longer required. Still, this requires the public to pay more now, and this approach is unlikely to be acceptable.

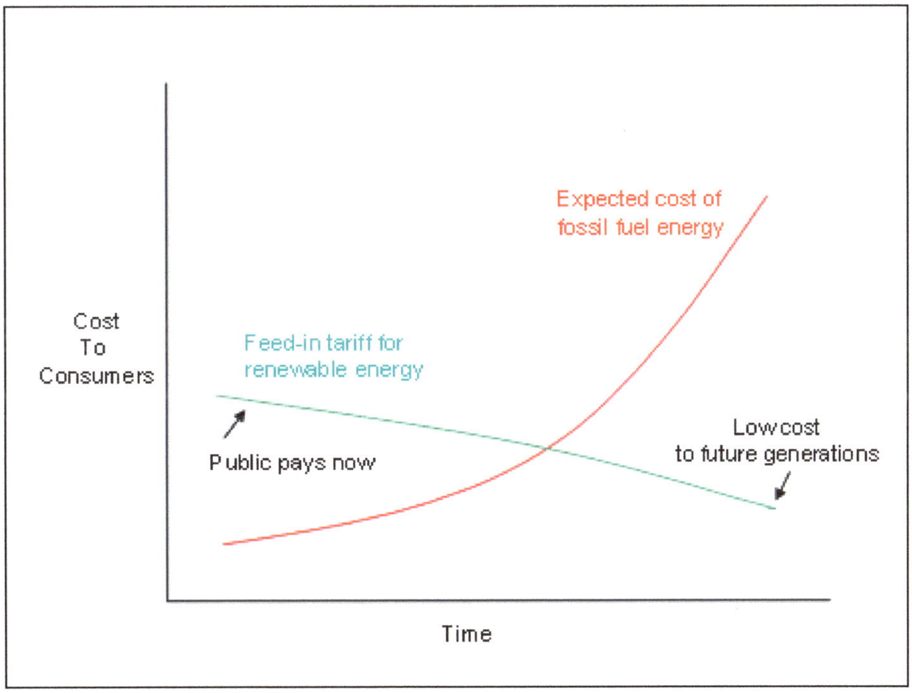

A different mechanism to required, where both current and future generations pay the costs. This is because the current generation will be unwilling to pay, even though rapid action saves money over the longer term.

The cost of 100% renewable energy implementation for Australia is large, at several hundred billion dollars. However, my letter of 19 April showed the cumulative upfront investment cost reaches a maximum of only $10B to $20B during the first 10 years, compared to business as usual, after coal savings are deducted. The eventual savings are in the order of one to two trillion dollars, in years 20-30, when the work becomes profitable compared to business as usual.

It would be immoral to miss out the future benefits, and instead incur the future costs, because we are unwilling to pay now for the small upfront investment that is required to commence the process.

The benefits to future generations are massive, even just in terms of reduced electricity generation running costs compared to business as usual. It could therefore be argued that it is reasonable for future generations to share the cost, rather than requiring it all to be paid now.

The obvious answer is therefore that the benefits should be achieved through borrowing, in order to smooth and share the cost to both current and future generations.

I see this approach as a solution to allow large-scale renewable energy implementation to commence now, whilst not requiring the current generation to fund the whole cost, with future generations then paying nearly nothing for ongoing cheap power.

The purpose of this borrowing is to ease costs to current consumers in paying feed-in tariffs to renewable energy generators. It is additional to any borrowing undertaken by renewable energy generators themselves as part of normal commercial operations, and then paid off by these tariffs.

The amount of borrowing can be adjusted to tailor how costs are shared between current and future generations. As one scenario, all additional funds would be borrowed and the current generation would not pay anything extra now at all. Given the expected benefits to future generations, rather than current ones, this would be reasonable. The upfront investment cost of $10B to $20B is small compared to current total national borrowings.

A sharing of costs between current and future generations would be more appropriate. However, the current generation should not need to be held or feel fully responsible for funding the cost of urgently replacing technology that has been established over the past two hundred years. We are however fully responsible for properly starting the process.

Conclusion and Summary

My book "Climate Change Explored and Explained" summarises climate science and policies. It describes why carbon pricing alone is unlikely to be efficient or effective. It provides a plan for Australia and other nations to implement 100% renewable energy and become carbon neutral across all emissions sectors in 30 years.

The cost of this plan for Australia is $13B per annum or $390B over 30 years.

My first update letter of 19 April 2011 shows that the savings for Australia from reduced coal usage for electricity generation are likely to exceed the cost of the plan by one to two trillion dollars. The analysis also shows that delay costs more, because we can expect the cost of fossil fuel in three decades time will dwarf the costs of renewable energy now. It shows that although renewable energy is currently more expensive, an aggressive renewable energy implementation plan is not.

This second update letter describes a means of funding the plan. Although the plan is profitable, the analysis indicates an upfront investment cost of a mere $10B to $20B over 10 years is required, depending on coal price assumptions. If mechanisms such as feed-in tariffs are used now to fund the current higher cost of large-scale renewable energy, this funding plan suggests that some or all of this additional cost could be borrowed in order to reduce or eliminate impact on our current generation.

Future generations will benefit enormously from embarking on this work now, with massive benefits of zero cost for fuel for electricity, transport and energy export, and a much reduced reliance on fossil fuels. They will easily be able to repay the very small upfront cost described above. In fact, the upfront investment cost is so small compared to current national borrowings that it would barely be noticed by any generation, future or current.

As my work and the work of Beyond Zero Emissions both show, the economic and risk management benefits to Australia of starting now on such a plan are massive, with benefits far exceeding costs. Setting an inadequate carbon price with no further mechanism to fund renewable energy is a waste of money and time. We might have money to waste but we do not have time.

Emission reduction targets of only 5% below 1990 levels by 2020 are inadequate, based on what the science says is required to limit our world to only 2 degrees of warming. Far more aggressive targets are thus required. We really need to return our climate to pre-warming levels, as melting icecaps, thawing tundra and a decaying Amazon are occurring now, and we don't know where this will lead.

To not start aggressively now runs the risk of future economic and environmental collapse. The costs of starting aggressively now are very small. Given the enormous risks, the low costs and the massive benefits, not starting aggressively now is irresponsible management of our economy, environment and way of life.

I would be pleased to discuss this further.

Yours sincerely

Stephen Harvie

References

"The Oil Crunch", ABC Catalyst, 28 April 2011 - An inability for world oil supply to meets demands is likely to rapidly push up oil prices in the near future. Apart from the impact on the price of petrol, this could have devastating impacts on world economies.

"Centennial Ready to Resume Coal Talks" Sydney Morning Herald, 25 April 2011 - Coal mining companies in Australia are currently seeking new contacts for coal for electricity usage that are closer to current market prices.

"Greens renewable electricity plan sparks power price fears", The Australian, 6 May 2011 - The electricity industry has warned that a Greens push to use the carbon-pricing debate to leverage a massive boost in renewable energy developments will dramatically increase electricity prices. Opposition climate action spokesman Greg Hunt said Senator Milne's proposal would double or triple power prices.

Attachments

A depiction of possible Australian and world political, economic and environmental futures.

A selected page from the 30 year renewable energy implementation plan – costs and savings based on a $120 per ton starting coal price (as per second letter). A number of different coal price scenarios were analysed, at $10, $50 and $120 per ton. The price is currently under negotiation, as discussed above.

Important Note

A detailed renewable energy implementation plan, such as the draft one attached, is required if the costs, benefits and means of funding such a plan are to be properly analysed and understood.

The Greens advocate the creation of such a plan by Infrastructure Australia. If that is agreed, that is good. It would be better if such a plan existed now.

The lack of an agreed plan currently hampers proper understanding and debate, and could prevent Australia from taking the best course.

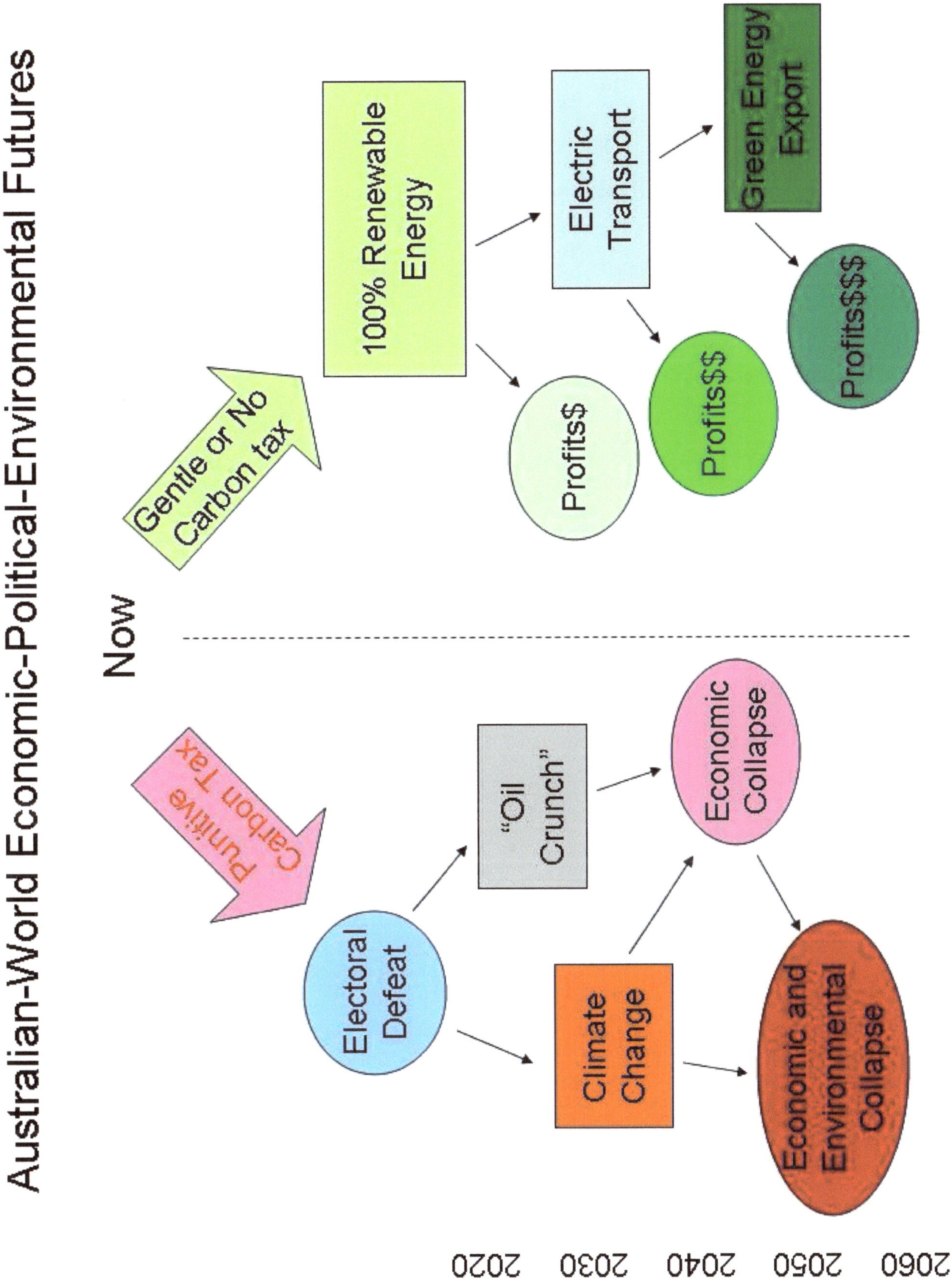
Australian-World Economic-Political-Environmental Futures
Now
Punitive Carbon Tax
Electoral Defeat
"Oil Crunch"
Climate Change
Economic Collapse
Economic and Environmental Collapse
Gentle or No Carbon tax
100% Renewable Energy
Profits$
Electric Transport
Profits$$
Green Energy Export
Profits$$$
2020
2030
2040
2050
2060

11 May 2011 Third letter

Senator the Hon Christine Milne
Deputy Leader of the Australian Greens
GPO Box 896
Hobart Tasmania 7001

The Greens Urgently Need to Become Australia's Sound Economic Managers

Dear Ms Milne

I write to you on this third occasion because I believe you are the only federal politician that talks sense!

I celebrated your announcements of last week (5 May 2011) that explained that a carbon price alone will not be effective and that a plan for 100% renewable energy is also required. I acknowledge that this is a continuation of the Greens' *SafeClimate* policies that have existed for quite some time.

I write to you particularly on this third occasion, because after reflection, I wish to make the following nine assertions and suggestions for improvement to your policy approach which I believe are urgently required.

1. Your argument that a plan for 100% renewable energy is required as well as having a carbon price is correct.
2. This argument is being denigrated by media and industry people as putting additional pressure on electricity prices e.g. "Greens renewable electricity plan sparks power price fears", The Australian, 6 May 2011.
3. I think both the carbon tax and your additional requirements described above are likely to be rejected by voters, as they will be seen to excessively hurt the hip-pocket.
4. I believe your arguments make the best economic sense but are not being sufficiently substantiated and explained.
5. I suggest the Greens will always be regarded as extremists, and not sound economic managers, until these arguments are more fully substantiated and explained.
6. The policy approach of requiring Infrastructure Australia to create the necessary renewable energy implementation plans now seems wrong from a timing perspective in the current political climate. Until the public and industry can be made to understand and believe in the financial benefits of an aggressive implementation of renewable energy, they will not vote for it or otherwise support it. In that case, the minority Government is also unlikely to support it.
7. The 30 year plans sent to you by myself (19 April 2011) and also the 10 year plan produced by *Beyond Zero Emissions* are first attempts to show the economic benefits and profits of rapid renewable energy deployment. Claims by the coal industry and others that a high carbon price is required to subsidise renewable energy will continue to hide and thwart what I think is the real truth. I discuss this more below.
8. Rather than requiring plans be created by Infrastructure Australia, the Greens need to urgently hire the experts, economists and marketing assistance required to substantiate the truth and convey the message, or otherwise face policy and/or election defeat. A very substantial television campaign would be appropriate.
9. The real need for a carbon price or emissions trading also needs to be urgently reviewed, from economic and political perspectives. My letter to you of 19 April explains why a carbon price is not needed to adopt a strong and profitable renewable energy plan.

I do not know if there will be an election over the issue of a carbon tax or emissions trading, or whether the Government will proceed with a carbon pricing policy in the context of your additional requirements, plus only modest public and industry support. In any case, obtaining support for a strong renewable energy plan will require much stronger presentation of the costs and benefits than is being provided at present.

Arguments for a Renewable Energy Plan

My letter to you of 19 April 2011 provided an analysis of the costs and financial benefits as an important update to my book "Climate Change Explored and Explained". This update highlighted that financial savings for Australia of between $278B and $2,268B over the next 30-40 years are possible from a rapid implementation of renewable energy, commencing now. With growing coal savings, the costs of the plan are not high at any stage of deployment, are easily met by any Government, and the plan quickly becomes profitable.

I believe I now understand the key source of public and industry confusion over the costs and benefits of renewable energy. Advocates of either coal or renewable energy, on either side of the fence, will both commonly argue that a carbon price of around $100 per tonne would currently be required to make the newer technology competitive. My Levelised Energy Cost calculations confirm this is true, based on current solar thermal costs of around $10.5M/MW.

However my analysis, and also that of *Beyond Zero Emissions*, shows that a 30 year or 10 year implementation plan based on a mix of thermal solar and wind power, with some price reduction expected over time, has an average Levelised Energy Cost that is comparable or even profitable compared to coal now. My plan shows an average $3.0M/MW capital cost of renewable energy, which is close to coal, but with no ongoing fuel costs. It also shows that delayed implementation costs more.

These comparisons show that implementation should commence now, as you have argued, even if only to save us money. These arguments can be very clearly presented, as demonstrated in the work of *Beyond Zero Emissions*. As one addition to the argument, neither of the analyses referred to above take into account the potentially large fossil fuel price rises resulting from peak oil. Starting implementation now could save us from a future crippled oil-based economy.

To base any arguments and decisions now only on price differentials at this very moment in time, and a starting carbon price, as people do, is not at all smart thinking. Renewable energy costs more now, but a renewable energy implementation plan does not.

Arguments against a Carbon Price

Further arguments can and will be made about the effectiveness and wisdom of carbon pricing.

> Firstly, a carbon price that starts low now is ineffective, so why have it?
>
> Secondly, a carbon price that is high later is not likely to be required due to the future price of fossil fuels, so why have it?
>
> Thirdly, the price differential between fossil fuels and renewable energy will reduce over time, and not increase, unlike a rising carbon price which works the wrong way around.
>
> Fourthly, my analysis and that of Beyond Zero Emissions both demonstrate that a carbon price is not required to make a renewable energy implementation plan cost competitive, or profitable.
>
> Fifthly, it is politically difficult.

From a renewable energy perspective, the carbon pricing approach is thus flawed. Your call for a renewable energy plan in addition to a carbon price reflects that. If we must have a carbon price, perhaps to encourage emission reductions in other sectors, one good approach is to make it very low.

Conclusion

The public is smart enough to recognise a good policy when they see one. They don't see one at present. I believe that the public and industry would understand these kinds of arguments, even if the ones you wish to pursue are only those concerning renewable energy implementation, rather than additional ones I make above, and further in my book, about the merits or otherwise of a carbon price.

A detailed renewable energy implementation plan of some nature is required now, not later, in order to demonstrate its expected financial benefits. One way to obtain a strong plan quickly could be for you to approach *Beyond Zero Emissions*. Although I do think their 10 year plan is possibly achievable, for increased credibility, I also think a modified version of their plan with a more flexible timeframe and other more conservative assumptions to suit your purposes could be quickly prepared.

I am considering as my next step approaching lobby groups such as *Get Up*. If you would have other information or better suggestions for me on how I might best proceed, I would be pleased to hear from you.

Yours sincerely

Stephen Harvie

References

"The Oil Crunch", ABC Catalyst, 28 April 2011 - An inability for world oil supply to meets demands is likely to rapidly push up oil prices in the near future. Apart from the impact on the price of petrol, this could have devastating impacts on world economies.

"Centennial Ready to Resume Coal Talks" Sydney Morning Herald, 25 April 2011 - Coal mining companies in Australia are currently seeking new contacts for coal for electricity usage that are closer to current market prices.

"Greens renewable electricity plan sparks power price fears", The Australian, 6 May 2011 - The electricity industry has warned that a Greens push to use the carbon-pricing debate to leverage a massive boost in renewable energy developments will dramatically increase electricity prices. Opposition climate action spokesman Greg Hunt said Senator Milne's proposal would double or triple power prices.

Attachments

A depiction of possible Australian and world political, economic and environmental futures (as per previous letter)

A selected page from the 30 year renewable energy plan (As per second letter - $120/ton coal price)

Fifth Letter

24 May 2011

The Hon Wayne Swan MP
Deputy Prime Minister; Treasurer
PO Box 6022
House of Representatives
Parliament House
Canberra ACT 2600

A Much Better Climate Policy for Australia

Dear Mr Swan

Please find enclosed the latest edition of my book about climate science and policies. It identifies:

1. **Savings to Australia of approximately $2 Trillion** over 30-40 years from **reduced coal usage** that can be gained from a rapid deployment of renewable energy, without delay.
2. How to fund the upfront additional investment to start the deployment, of only $1.5B p.a. for the first 10 years, **at little or no additional cost to consumers**, until the profits begin.
3. A plan to use a portion of these massive profits to move **Australia to carbon neutral** across all emissions sectors, as well as implementing renewable energy, in only 30 years.

This assessment adds to other excellent work undertaken by *Beyond Zero Emissions* in their 10 year renewable energy implementation plan. Their plan also identifies **$1 Trillion of savings** over 30 years from **reduced oil usage** through electrification of public transport.

A further **$1 Trillion of savings** can also be achieved over a similar timeframe through accelerated conversion of private transport to electric cars, as proposed in my book.

This totals **$4 Trillion of direct savings** to Australia over 30-40 years from rapid renewable energy deployment. Even more significantly, it also provides Australia with energy security and protection against **wider economic impacts** of rising fossil fuel prices, as well as climate change.

The funding and implementation method described in my book is superior to carbon pricing, which is in comparison inefficient, ineffective, slow, expensive and unpopular as a result. I see Canada has now dropped discussion of carbon pricing after many years of debate. My book suggests how to fund an aggressive renewable energy plan so that a carbon price can be avoided or kept small.

Interestingly, my book estimates that $4 Trillion is also the amount required to convert the USA to renewable energy. This paves the way for possibly the biggest commercial deal done by mankind, if we were to create the necessary partnership. The potential for this is also discussed in my book.

Most of the necessary supporting information for the figures above is provided in the update letters included as the first 20 pages of the enclosed book. A summary is also attached. I would be pleased to discuss these figures further.

Yours sincerely

Stephen Harvie

Fifth Letter

30 May 2011

The Hon Joe Hockey MP
Shadow Treasurer
PO Box 6022
House of Representatives
Parliament House
Canberra ACT 2600

A Much Better Climate Policy for Australia

Dear Mr Hockey

Please find enclosed the latest edition of my book about climate science and policies. It identifies:

1. **Savings to Australia of approximately $2 Trillion** over 30-40 years from **reduced coal usage** that can be gained from a rapid deployment of renewable energy, without delay.
2. How to fund the upfront additional investment to start the deployment, of only $1.5B p.a. for the first 10 years, **at little or no additional cost to consumers**, until the profits begin.
3. A plan to use a portion of these massive profits to move **Australia to carbon neutral** across all emissions sectors, as well as implementing renewable energy, in only 30 years.

This assessment adds to other excellent work undertaken by *Beyond Zero Emissions* in their 10 year renewable energy implementation plan. Their plan also identifies **$1 Trillion of savings** over 30 years from **reduced oil usage** through electrification of public transport.

A further **$1 Trillion of savings** can also be achieved over a similar timeframe through accelerated conversion of private transport to electric cars, as proposed in my book.

This totals **$4 Trillion of direct savings** to Australia over 30-40 years from rapid renewable energy deployment. Even more significantly, it also provides Australia with energy security and protection against **wider economic impacts** of rising fossil fuel prices, as well as climate change.

The funding and implementation method described in my book is superior to carbon pricing, which is in comparison inefficient, ineffective, slow, expensive and unpopular as a result. I see Canada has now dropped discussion of carbon pricing after many years of debate. My book suggests how to fund an aggressive renewable energy plan so that a carbon price can be avoided or kept small.

Interestingly, my book estimates that $4 Trillion is also the amount required to convert the USA to renewable energy. This paves the way for possibly the biggest commercial deal done by mankind, if we were to create the necessary partnership. The potential for this is also discussed in my book.

Most of the necessary supporting information for the figures above is provided in the update letters included as the first 20 pages of the enclosed book. A summary is also attached. I would be pleased to discuss these figures further.

Yours sincerely

Stephen Harvie

Summary of $4 Trillion Savings to Australia over 30 - 40 years from Rapid Renewable Energy Deployment

$2 Trillion from reduced coal usage	This is explained in detail in *Climate Change Explored and Explained* update letter of 19 April 2011 – ***Financial Savings of Rapid Renewable Energy Implementation and Potential for a Low Carbon Price***. The total proposed renewable energy capital implementation costs are towards $10B p.a. for 30 years. The costs grow progressively in the initial years. These costs are offset by elimination of new coal fired power station capital costs and also a growing elimination of coal fuel costs. As large-scale renewable energy is rolled out, these eliminated coal fuel costs also grow progressively towards $10B p.a. in the first 10 years and to over $100B p.a. in years 30 to 40. The net cost reduces to around $1.5B per annum in years 1 to 10 and then the plan becomes increasingly profitable over years 20-40.
$1 Trillion from electrification of public transport	Reduced oil usage over 30 years as estimated by *Beyond Zero Emissions* in their *Zero Carbon Australia 2020 Stationary Energy Plan*.
$1 Trillion from electrification of private transport	See *Climate Change Explored and Explained* draft climate policy (page 41) – Green Car Subsidy. Profits from large scale renewable energy deployment can be used to subsidise 5 million new private electric cars, which is one third of Australia's privately registered vehicle fleet – estimated policy cost $75B over 30 years. If reduced petrol usage is $5,000 per car per annum (a conservative estimate of future petrol prices), over 40 years the total saving of 5 million additional electric cars on the road, powered by renewable energy, is $1 Trillion.
Total $4 Trillion	Over 30-40 years, from reduced coal usage for electricity generation, and oil usage for public and private transport.

For a discussion on how to fund renewable energy implementation at little or no current cost to consumers, see *Climate Change Explored and Explained* update letter of 18 May 2011 – ***The Potential Climate Policy Impasse, How Current and Future Generations Should Pay for Renewable Energy***.

Book Website - Home Page - September 2011

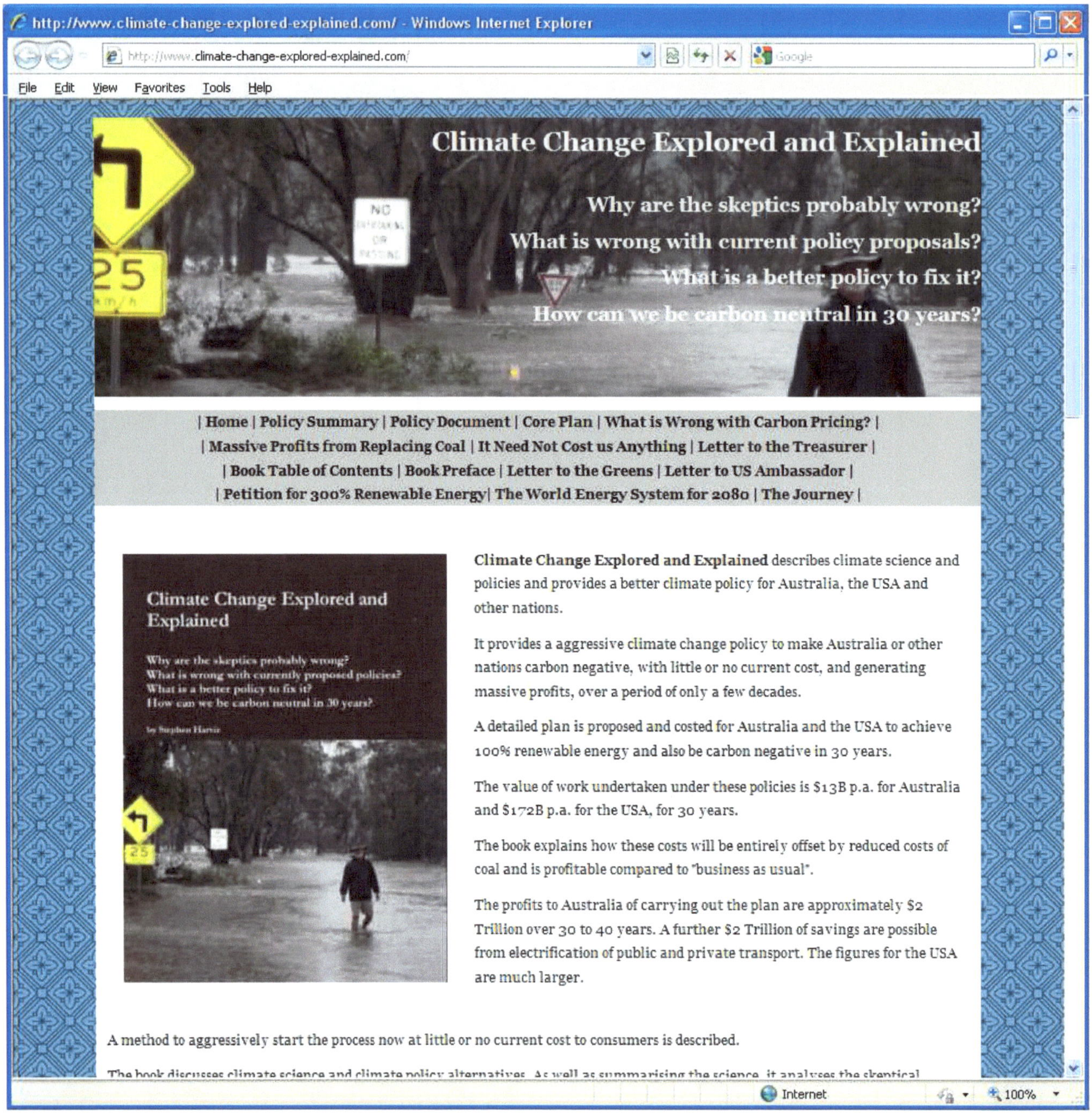

The following text is a copy of the material from the web site home page and then each of the tabs at this website. It includes analysis of current government policies, better proposals, and proposed costings and profits for Australia to provide 10% of the world's electricity needs with renewable energy.

Climate Change Explored and Explained describes climate science and policies and provides a better climate policy for Australia, the USA and other nations.

It provides a aggressive climate change policy to make Australia or other nations carbon negative, with little or no current cost, and generating massive profits, over a period of only a few decades.

A detailed plan is proposed and costed for Australia and the USA to achieve 100% renewable energy and also be carbon negative in 30 years.

The value of work undertaken under these policies is $13B p.a. for Australia and $172B p.a. for the USA, for 30 years.

The book explains how these costs will be entirely offset by reduced costs of coal and is profitable compared to "business as usual".

The profits to Australia of carrying out the plan are approximately $2 Trillion over 30 to 40 years. A further $2 Trillion of savings are possible from electrification of public and private transport. The figures for the USA are much larger.

A method to aggressively start the process now at little or no current cost to consumers is described.

The book discusses climate science and climate policy alternatives. As well as summarising the science, it analyses the skeptical arguments and describes why these are probably wrong. The book discusses policies for the world, and those proposed in the USA, but focuses particularly on the alternative emissions trading, carbon tax and direct action policies currently proposed in Australia.

The book describes why carbon pricing approaches using emissions trading or a carbon tax might not work. Current emission reduction targets are not thought to give us a 50% chance of avoiding dangerous climate change. This book proposes a very large-scale direct-action plan to solve the problem of climate change.

A response to the Australian Government's newly announced carbon pricing package is also now provided at this web site at tabs "**Policy Summary**" and "**Policy Document**". These explain why emissions trading is far less efficient and effective than this book's policy proposal, being slower, less effective, and less efficient at approximately three times the cost.

The analysis of costs of implementing 100% renewable energy costs and the massive savings for Australia over a 30-40 year period are available by contacing the author at the email address at the back of this book. The equivalent USA analysis is also available. The purpose of sharing these openly is to encourage wider analysis, verification and improvement.

Once you make the mental leap that very large scale renewable energy is cheaper than staying with coal, you can then start to understand why an ineffective carbon price, or feed-in tariffs that make consumers pay more now, are not the most effective ways to go.

Some of the key updates to the book since its first July 2010 edition are shown as tabs on this web site.

The **Massive Profits from Replacing Coal** update letter (second letter - front of book) explains savings from coal usage exceed the costs of the plan to quickly implement 100% renewable energy.

The **It Need Not Cost Us Anything** update letter (fourth letter - back of book) explains how the small initial investment to start the plan (of only $10B-$20B for Australia) can be spread across current and future generations. The massive profits, benefits and cost advantages will be mostly experienced by future generations.

The **Letter to the Treasurer** (fifth letter - front of book) was sent to the Australian Treasurer and Shadow Treasurer to endeavour to interest them in the book's policy by explaining the significant profits of the proposed plan, at several trillion dollars. It is now the new covering letter in the book.

The **Petition for 300% Renewable Energy** proposes a better way for Australia to spend the funds to be sent overseas from emissions trading.

The **World Energy System for 2080** provides a renewable energy solution for the entire world. It also describes how Australia could profit from supplying 10% of the world's electricity with renewable energy, at the same cost as the emissions trading proposal. This equates to a bit over 600% renewable energy, would generate massive profits and make Australia several hundred percent carbon negative.

How in 50 Years Australia Could Supply 10% of the World's Electricity Needs with Renewable Energy

- Australia's total land area is 7,687,000 square kilometres - 5% of all land on earth.
- 225 square kilometres is the size of one patch of land 15 kilometres by 15 kilometres that would hold 10 large circular 220MW solar thermal power stations.
- If we have 30 regional or remote sites that could accomodate such a patch of land, on all 4 sides, this is space to build over 50 years 30 X 4 X 10 = 1200 power stations.
- Each of these stations produces 1400 gigawatt hours of electricity per year, based on typical weather patterns. The total power generated from 1200 stations is 1700 terawatt hours per year, or 10% of the current world electricity usage.
- The land are used is 225 X 30 X 4 = 27,000 square kilometres, or only 0.35% of Australia's total land area (one third of 1%).
- To meet this target of 1200 stations in 50 years we need to build 24 per year. It is a large but achievable target, particularly if we start now.
- At $2.2 Billion each, the total cost of building 1200 power stations over 50 years is about $50 Billion per year or $2.6 Trillion in total.
- This cost is high but the potential profit is huge.
- The potential profit over the first 50 years is $77 Trillion. This is about 70 times Australia's current annual GDP. This estimate is based on wholesale electricity prices starting at 5 cents per kilowatt hour and rising at 10% each year for 50 years, to around $5 per kilowatt hour in 2050. It is the profit initially from onshore usage and then energy export over 50 years as the system is gradually built. Income continues after that at potentially $9 Trillion per year (1700 terawatt hours at $5 per kilowatt hour in 2050) which is about 9 times our current annual GDP.

The **World Energy System for 2080** tab provides details of this plan for a 100% renewable energy system for the world and the significant role Australia can play in this.

Comments about the first edition of the book from current prominent Australian politicians from each of the major political parties were that it is "very valuable", "of great interest", "comprehensive", "well-researched", "well-argued", "valuable in informing the development of Australia's foreign policy" and that "the policy proposals and suggestions for action are informative for anyone who is seriously concerned about the future of mankind and our planet".

Also another comment recently received is "A truly spendid contribution to the climate change debate".

Revenue from books sold on-line goes towards recouping substantial costs of the 50-60 books that have been sent to politicians, web site costs and legal advice fees.

I also highly recommend the Australian **Beyond Zero Emissions** Renewable Energy 10 year implementation plan. It is also a must read and probably the best work of its type in the world.

Policy Summary

Below is a summary of the proposed policy, using the Australian cost and benefit figures, as well as a comparison with emissions trading.

The detail of the cost-benefit analysis figures for Australia are presented in letters at the front of this book and available, as well as details for the USA, by contacting the author.

A Better Climate Plan for Australia

1. **100% renewable energy for Australia** will cost $300 Billion, based on current technology that is operational in Spain and in construction in California. Using large-scale energy storage, renewable energy can meet Australia's baseload and peak energy requirements. The proposed implementation time is 30 years. A huge number of new green jobs will be created. A further $100B can get Australia to carbon negative in the same timeframe, funding measures across transportation, agriculture and international reforestation.

2. The **savings** from reduced coal usage, at between $1-2 Trillion over 30-40 years, exceed the cost of the plan, meaning that it is profitable. This is because of rising fossil fuel prices. If we want, we can continue to mine and export our unused coal, while other countries still need it.

3. **Further saving**s of $2 Trillion over the same time period can be achieved through electrification of public and private transport, through reduced oil usage. The plan includes subsidies for 5 million electric cars.

4. A large number of **jobs** are created.

5. Although the cost of capital work for Australia exceeds $10B per annum, the upfront investment in the first 10 years is only $20B, after savings from reduced coal usage and no new coal power plants is deducted. After that, the massive profits of the plan kick in at an increasing rate in years 10 onwards (see tab The Massive Profits from Replacing Coal for details). The very small upfront investment, of less than $2B p.a. for the first 10 years, should be borrowed so that consumers pay nothing more for the plan now (see tab/letter It Need Not Cost Us Anything). Future generations will easily repay the small debt when they benefit from 100% renewable energy.

6. By comparison, the **emissions trading** proposal is **slow, expensive and inefficient**. The total cost through to 2050 will exceed $1 Trillion, and the policy achieves much less effect. The additional cost results from the inefficiency of converting to gas power first, and from sending much of the funding (approx. $780B over 35 years to 2050) to uncoordinated carbon offset schemes overseas. A liability of $58B per year continues after 2050 to continue offsetting our onshore emissions. At Treasury's higher prediction of the international carbon price at 2050 of $275/ton, the total amount sent overseas then becomes $1.57 Trillion, taking the total emissions trading cost to $2-3 Trillion (see tab Policy Document for details). Funds sent overseas might not produce real emission cuts.

7. The **emissions trading** proposal achieves next to no net emission reductions onshore and creates an ongoing need to **send money offshore** (see Australian Treasury modeling below). The carbon price will not, for a long time, fund renewable energy, with only 40% predicted in 2050. The Treasury core policy modeling of an onshore reduction of only 2% from year 2000 emissions and only 5% solar energy by 2050 is a shameful vision for Australia. The proposed green energy fund is a very small offer of loans only - the carbon price is not sufficient to create the necessary investment.

8. For the same cost as emissions trading, of over $1 Trillion, and possibly over $2 Trillion, we could build renewable energy to meet **400% of Australia's total energy needs**, and **replace coal with green energy exports**, which is indeed possible to do. Even with only the money proposed to be sent overseas, we could achieve between 200% and 400% renewable energy. Europe is planning to source its solar power from the Sahara. Australia could service Asia. The sun is the new oil, and with the 2nd highest solar irradiance in the world, Australia could be a very rich country.

Expected Outcomes from Emissions Trading - Treasury Modeling showing actual onshore emissions reductions of only 2% in 2050 and abatement sourced from overseas (blue)

See Tab Petition for 300% Renewable Energy for the Total Expected Cost of Overseas Permits

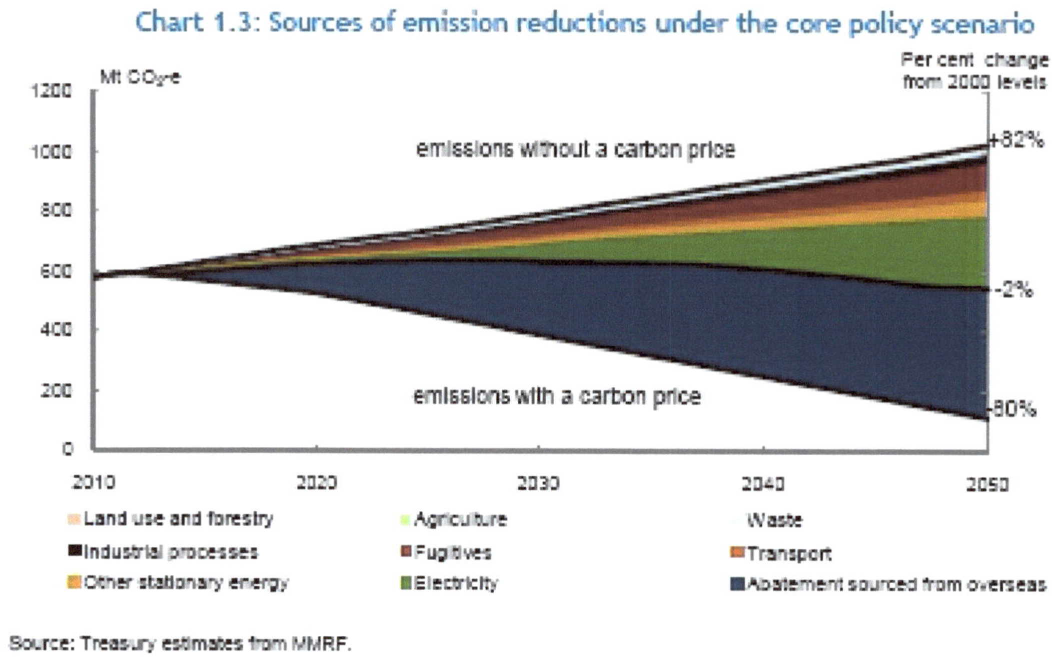

Source: Strong growth, low pollution - modeling a carbon price, The Treasury, Australian Government, 2011

Policy Document

Climate Change Explored and Explained

A Better Climate Plan for Australia

Policy Launched 10 July 2011

The following is a more detail summary document describing the large-scale direct-action climate policy proposal contained in this book. Also provided is a critique of the Australian Government's new carbon pricing proposal, which was launched on the same day.

100% Renewable Energy in 30 Years

A plan for 100% renewable energy is costed for Australia, the USA, China and the world.

A 100% renewable energy system for Australia will consist of a mix of solar thermal, wind, biomass backup and existing hydro-electric schemes. Combined with solar thermal energy storage systems, this mix can meet our full baseload and peak energy requirements.

An implementation timeframe of 30 years is proposed, which requires us start aggressive implementation now.

For Australia, the capital cost is around $300B, or $10B pa for 30 years. For the USA, the capital cost is around $4 Trillion, or $133B pa for 30 years.

Savings from Reduced Coal Exceed the Costs of the Plan

The savings from reduced coal usage for electricity generation exceed the cost of the plan by a very large extent.

The book calculates a profit, or savings, for Australia from undertaking the above plan at between $1 Trillion and $2 Trillion dollars, over a 30-40 year timeframe.

The profit for the USA is much larger.

Detailed costings for the implementation plan and the expected savings from coal usage are provided in the book and available by contacting the author.

Ways to fund the Plan

There are several ways to fund the plan.

If we ignore the savings from reduced coal usage, two methods could be:

1. Based on Australia's current energy usage of 257Twh p.a, a simple levy of 5 cents per kilowatt hour would provide full funding for the plan.

2. A more complex way would be to have a carbon tax, and use the revenue to fight climate change, rather than to provide compensating subsidies.

A better funding method is proposed in the book. Although the capital cost for Australia is $10B per annum, after coal savings are deducted, only a mere $10-$20 billion dollars needs to be borrowed in the first 10 years to commence building the large-scale facilities. After that, the savings begin and costs of the plan are much less than staying with coal (details of these costings are provided in update letter two.)

Such borrowings are very small compared to current national borrowing, is only about half the cost of the National Broadband Network, is equal to only about half of one year's Defence expenditure, and only represent 0.1-0.2% of GDP per annum

These borrowings can be used to provide feed-in tariffs to fund the facilities, so that the public need not pay anything extra now (as explained in update letter four).

For future generations, the running costs of renewable energy will be close to zero, while fossil fuels skyrocket. Future generations will then have much lower energy costs, and will easily repay these very small upfront borrowings required to commence the plan.

This creates a better model for "inter-generational equity", where the current generation does not have to pay the full costs, and future generations receive the massive economic and environmental benefits of starting the plan now.

The book shows that delay costs more, because rising fossil fuel prices will dwarf falling renewable energy prices. Although renewable energy is more expensive now, to delay implementation is more costly over a 30-40 year time period than starting aggressively now. The massive profits start after year 10.

The current generation need not be held fully responsible for funding the very small upfront cost of commencing the plan. We are however fully responsible for starting the process. In this regard, we are failing.

Getting to Carbon Negative

The book costs a plan to not only implement renewable energy, but also to get Australia, or other nations, to carbon negative within 30 years.

For Australia, the additional cost is approximately $100B over 30 years, which a small addition to the plan and well within the savings from reduced coal usage.

For the USA, the additional cost is $1.3T, but again, this is well within the savings from reduced coal usage.

It is actually very hard to do, without looking to opportunities off-shore. It is impossible to eliminate all emissions from industrial and household processes. The book suggests measures that can be taken particularly across sectors of transportation and agriculture.

In the transportation section, for Australia there is potential for a further $2 trillion of savings over 40 years from electrification of public and private transport. The book's policy proposal costs subsidies for 5 million electric cars for Australia and 50 million for the USA. Under the book's plan, these will run on renewable energy.

The plan to get to carbon negative allocates funds, ideally as part of an international effort, to eliminate deforestation of rain forests across eight key foreign nations, which is an important opportunity for direct-action identified in the Stern review. This is thought to currently account for 20% of the world's emissions, and thus must be stopped.

The detailed plan that is provided in the book, for both Australia and the USA, shows how both nations can get to a significantly carbon negative position. It is also an approach required on moral grounds, is effective, and is financially profitable.

The plan effectively addresses the environmental, economic and even more significant risks of climate change, as well as the risks of energy security and of potential economic collapse resulting from peak oil.

Achieving International Action

There is debate in Australia about whether a small country such as Australia should take action at all.

Yes we should. Australia is one of the world's highest per capita emitters and we cannot morally maintain that position. Australia is also a country forecast to be amongst the first nations to collapse from potential effects of climate change on our economy, food supply and environment.

Further, we are the world's 14th highest emitter in total. If we don't need to act, there are over 200 countries behind us that would not need to act either, counting for a large part of the world's emissions.

Australia has the world's 2nd highest solar radiation, after Mongolia. We are the country best positioned to show the world the best way forward, become the world's leading solar thermal supplier. China has taken over our photovoltaics. After we have met our own energy needs, we can then start to export renewable energy to Asia, or other parts of the world, rather than coal. That is indeed possible. Europe is planning to source its renewable energy from the Sahara.

Australia has the opportunity to undertake a plan that is not only profitable for ourselves, but shows the world what it needs to do, to avoid the significant risks of climate change. One does not need to accept the scientific concerns for the future our planet, and our health towards the end of this century, for this to be a good economic plan. Again, as an added benefit, the plan also provides for energy security, and addresses risks from economic collapse resulting from peak oil.

The State of the Policy Debate

The state of the policy debate in Australia is substandard. There is no real analysis of whether measures such as a carbon tax, or emissions trading, are really effective or efficient measures to address the problem, and whether they can work in the required timeframe.

We hear only very simple claims of whether or not they are "efficient", based on what some experts say. It isn't clear what "efficient" really means.

Analyses of overseas carbon pricing schemes by environmental groups such as Friends of the Earth, and some economists, explain why carbon pricing is not efficient or effective and cannot work in the required timeframe.

Some of the key reasons carbon pricing is ineffective include:

1. The starting carbon price will be too low to fund renewable energy, and will barely be sufficient to even convert to gas, which provides little or no advantage, is therefore a waste of time and money (we don't have time to waste).

2. A carbon tax does not guarantee any level of action. Compensating subsidies to industry and consumers will mean no-one really has to do anything.

3. Emissions trading does not provide any certainty of carbon price, and so investment will be stunted. This is proven overseas.

4. The emissions trading scheme will allow up to 50% of investment to go overseas, and experience has shown that overseas offsets schemes in particular can result in emissions reductions not being real.

5. There are a range of "non-price barriers to action", including (1) split incentives between consumers and various parts of the supply chain; (2) difficulties in coordinating investment; (3) different investment priorities and constraints; (4) lack of awareness of options; and last but not least (5) uncertainty of future government policy. These are some of the barriers to action identified by ClimateWorks Australia, and will hold back real action. Climate mitigation is not core-business of industry.

These are the practical issues that the "economic experts" who advocate carbon pricing will generally ignore, are not properly debated, that we know will hold back real action, but which are hard for most of us to describe.

We are told that carbon pricing will promote investment in renewable energy. In reality, experience overseas shows that carbon pricing will prevent investment in renewable energy.

Carbon pricing can a useful part of a broader, more effective policy, but is not an effective policy by itself. A low carbon price, with subsidies, is indeed a gentle way to start to move towards a low carbon economy. However it does not provide the level of climate action that is required. Further, if it only funds conversion from coal to gas, it actually prevents the level of action that is required.

There are additional reasons why we are confused about renewable energy. We only hear about its costs now. There is no long term analysis and planning. Although the capital cost of renewable energy is more expensive than coal now, a 30 year renewable energy implementation plan is not. It is cheaper.

On television, discussion about solar energy usually shows pictures on panels being installed on roofs of houses. The Australian opposition party suggests a policy of solar panels on 1 million Australian roofs. No-one explains that 1 million roofs, at around 2KW each, and an average 1500 hours of sun each year, delivers a total of only 3 terrawatt hours a year, or only about 1% of

Australia current energy usage. Such a policy is costly and thus barely worth doing. A larger plan, such as the one described in this book, is required.

A comparison of this book's policy with the Australian government's new carbon pricing policy is provided below.

The most recent debate since the carbon pricing details announcement also remains substandard. The Australian Coalition clumsily points out that $58B will be sent overseas in 2050. They don't explain well that is the amount is posted away every year and probably continues to grow beyond that. They don't point out that the total over 35 years is $780B. They don't point out what else could be done with that money. They don't point out the total amount sent overseas becomes $1.57 Trillion at the Australian Treasury Department's higher prediction of the international carbon price at $275/ton. A link to an analysis spreadsheet is provided below.

The Australian Government simply counters that the economy will be three times bigger by then so the above figures are small and won't matter. That claim of economy growth could be hogwash for a start, and beyond that, if the number is so insignificant then why couldn't Australia use it to fund a better target for onshore renewable energy. Australia is proposing the biggest per capita emissions trading scheme in the world, and proposing to post much of the money overseas. The world will think of Australia as either the most generous country, or the dumbest country - take your pick. Some of the ways carbon offets can be spent overseas is on newer technology, lower emission, coal-fired power stations in India and China, or paying New Zealand to cut down less trees (they will chop them down later instead). Rather than send money to uncoordinated and doubtful overseas offset schemes, the world would thank us more if we spent it more effectively onshore.

Cost and Benefit Comparison of This Policy with the Australian Government's New Climate Policy Proposal

As well as its inefficiencies, other significant shortcomings of the Australian Government's New Climate Policy Proposal are as follows.

Inadequate emission reduction targets

The most disappointing thing about the Government's emissions trading proposal is that its targets are the minimum level thought to give the world only a 50% chance of avoiding 2 degrees of warming, a dangerous level of climate change. Temperature rises will be higher in Australia. How can we set such poor targets to protect our ongoing existence in this country, or on this planet?

Nations in the developed world are expected to do act faster, in order to meet a global carbon budget. Because of the complexity, problems and inefficiencies of emissions trading, we probably won't even achieve our targeted level of result.

Inadequate carbon prices

The proposed carbon prices will not fund renewable energy for a very long time. The Treasury models suggest $60 per ton carbon price will be required in order for funds to be wasted on conversion to gas. The targets set for renewable energy in Australia, at 40% by 2050, are less than those predicted to be achieved by the rest of the world.

Only 5% solar is expected in the Treasury models by 2050. Higher targets for dirty approaches such as gas, and hoping we can viably and safely pump CO_2 from coal into the ground, without endangering future generations, are also disappointing.

Poor onshore investment and emissions reductions

Under the core policy, there are no onshore emission reductions through to 2050. They remain at 600MT p.a. There is only an elimination of emissions growth. All reductions below that level are achieved by purchasing credits offshore, on an ongoing annual basis.

The pretence of supporting renewable energy

Many people are obviously excited by the proposed $10B clean energy fund. However only 50% of this will be directed to true renewable energy, and this total budget is provided over 5 years. That is only a mere $1B per year. Lastly, the main problem is that this is an offer of loans or loan guarantees only, if anyone wants to invest, rather than an actual intention to carry out any action.

What is the actual incentive for anyone to invest in renewable energy? It is not the carbon price - it isn't high enough. Some commentators have suggested the funds, or loan guarantees, won't be used.

Cost Comparison

The cost of the Australian emissions trading proposal is roughly as follows, based on figures in the Treasury documents (core policy option):

1. $200B estimated onshore investment in new electricity generation technology to 2050

2. A further unspecified amount of investment in other emissions areas, probably roughly equal to the above

3. $780B, being total investment sent overseas over 35 years. The annual amount is planned to grow until 2050, starting from zero to around 450 megatons of overseas sourced abatement, and with international carbon price predicted to grow from its 2015 starting point to $131 per ton. The multiplication of these two figures grows annually to $58B in 2050. The total over 35 years is $780B. The real cost could be much higher, with Treasury predicting international carbon prices could grow to higher levels of $275 per ton. At that level of growth, the money sent overseas over 35 years doubles to $1.57 Trillion (see details in a following analysis spreadsheet).

4. $58B per annum after 2050, predicted by Treasury to require 450 megatons of abatement per annum at $131 per ton in 2050, and this is ongoing. Depending on the international carbon price, the amount could be double this.

5. All the funding to middlemen, such as onshore and overseas banks that engage in the carbon trading, squirrel money away here and there, which is an amount which is incalculable but could easily represent half of the useful funding. Indeed, the carbon market can become much bigger than the amount of money itself. Financial markets support emissions trading.

All of these figures are in today's dollars. The amount spent onshore is probably an underestimate, given that, in theory the money sent overseas is for carbon abatement that is cheaper.

The proposed end point of the Treasury core emissions trading policy plan in 2050 is a nonsense, with only a 2% onshore emission reduction from current levels achieved, and all further reduction to achieve the 80% reduction occurring overseas. At that point we would more likely embark on the $390B direct action plan described in this book, rather than continue the debacle of sending $58B per annum of investment overseas.

That gives a total cost of the emissions trading proposal at well over $1 Trillion, being the above figures $200 + 200 + 780 + 390 billion in today's dollars. It is over $2 Trillion if the Australian Treasury's higher predicted 2050 carbon price of $275/ton is used.

Emissions trading is thus far less cost effective and efficient than this book's proposal. It is at least three times the cost, at $1-2 Trillion extra.

Should we care about an extra $1-2 Trillion over 35 years? At the lower end, it is about one National Broadband Network project each year for 35 years, which is a lot. At the lower end, it is enough funding to generate around 4 times our energy needs, which would create a significant green energy export business, replacing coal. Australia could be a rich country.

This book's policy, even at its lower targets of only 100% renewable energy, achieves a much better outcome, faster, cheaper, with much greater certainty. It is financially profitable, and and unlike the Government's proposal, aims to achieve a better than 50% chance of avoiding dangerous levels of climate change. Surely, this is what we want.

What is Wrong with Carbon Pricing?

Over the coming year Australia will need to take stock of its newly proposed carbon pricing policy. Most carbon pricing supporters would say that it is at least a step in the right direction. I disagree. I see it a a very inefficient and wasteful process that will prevent Australia from taking any real action, and gaining any significant environmental and economic advantages, for at least a decade or two. It will cause significant pain for very little gain. It holds back real climate change mitigation.

This report from Friends of the Earth, A Dangerous Obsession (5MB), describes why emissions trading is ineffective and fails. It would probably be a surprise to many Australians that many environmentalists don't support emissions trading. It has many problems that prevent real action on climate change. Low carbon prices, as well as price volatility and uncertainty all hold back real investment such as in renewable energy. Also purchasing carbon offsets allows pollution to continue often without any real emission cuts. More discussion of why emissions trading is ineffective is at Carbon Trading - a critical conversation on climate change, privatisation and power.

Why Does This Book's Policy Work Better than Emissions Trading?

The book's policy plan has the lowest cost and also the better outcome. If emissions trading looks for the least-cost option, why wouldn't emissions trading hunt out this option? This is a tough question to answer, but the book and material at this site provide some answers. In summary, some key reasons are:

1. **Emissions trading has many inefficiencies.** One is the split incentives, where industry need not do anything if it is easier to pass on costs to consumers. Another is that always choosing the least-cost option at the present moment, such as gas power, means the problem has to be solved over-and-over again in increasingly better ways. It is much cheaper to solve the problem once.

2. **Emissions trading lacks forward planning**. Under emissions trading, industry always takes the cheapest option for investment at that point in time. While coal is cheaper than renewables, that is what industry will prefer to stick with. The carbon price won't be high enough to make the necessary difference for a long time. A greater level of long term planning, such as the plan presented here, is required to realise that an immediate plan to convert to 100% renewable energy, with its 30 year lead time, is already cheaper than staying with coal, and no carbon price at all is required. Although renewable energy is currently more expensive that coal, an aggressive 30 year implementation plan is not - it is cheaper.

3. **Vested current interests,** especially in relation to fossil fuels, and particularly in Australia.

4. All of the other "**non-price barriers to action**" documented by ClimateWorks Australia, such as lack of awareness of solutions, split incentives (again), time delays in decision making, long payback periods, and last but not least for Australia's political climate, uncertainty of future prices and policy.

5. It is a slow, uncoordinated way of achieving emissions cuts, and particularly for funds sent to overseas offset schemes, it might not result in real cuts in emissions at all.

I was interested to read recently that our current Prime Minister was challenged with information from Deloitte that coal usage would not reduce until the carbon price gets to \$40-\$70 per ton. Apparently she said she is more optimistic. I also stumbled upon a 2001 USA study that shows that under three different electricity sector models, renewable energy actually reduces at carbon prices of \$25 and \$75 per ton. It explains this is because available investment all rushes to gas. If this is true, then our current carbon price advertisements are misleading. It is however great for Australia's Coal Seam Gas industry and all of its environmental and economic downsides.

Proposed Amendment to Australia's Emissions Trading Proposal

If we must have emissions trading, with all of its middlemen, rorting* and inefficiencies, then so be it. But we should set our sights higher.

* An Australian term.

Proposed Amendment - The $780B that would be sent over 35 years to overseas carbon offset schemes, will instead, for the same price, be spent on (1) building renewable energy to meet **200% of Australia's total energy needs***, for our own use plus green energy export, as well as (2) electrifying public and private transport, in order to protect Australia from the economic effects of peak oil. This will make Australia a rich country.

* Although for the full cost of emissions trading we could meet 400% of our energy needs, we unfortunately might need to settle in the short term for this less profitable outcome, using only the 50% overseas funds component, for political reasons. If the international carbon price rises to the higher predicted level of $275/ton, then we could achieve 400% renewable energy, still only using half of the emissions trading funding - the amount expected to be sent overseas.

This provides an overall solution which is 50% emissions trading, and 50% large-scale direct-action, which is an ingenious policy combination. It is the full emissions trading proposal but with significant leadership applied to the electricity sector. See tab **Petition for 300% Renewable Energy** for more discussion.

Further detail of the advantages of this book's policy over carbon pricing, and of broader problems in Australia's overall carbon pricing policy, is provided in tab Policy Document.

Other Commentary

Recommended further reading from Canada, a newspaper article available on the web titled..."To Work, Carbon Tax Must Sting".

It doesn't take a rocket scientist to work out that a weak carbon tax will do very little. As a country, Canada has rejected carbon pricing in favour of other methods.

Proposal for Petition for 300% Renewable Energy for Australia by 2050

I became aware of the potential for Australia to export renewable energy when I read the most wonderful plan for 100% renewable energy from Beyond Zero Emissions (you should download the plan from their web site).

Below is an email sent to some environmental groups regarding the start of a formal petition for 300% renewable energy for Australia by 2050.

Petition for 300% Renewable Energy

I wish to invite you to help organise a formal petition for 300% renewable energy for Australia by 2050.

100% renewable energy for Australia has been estimated to cost around $300 Billion by Beyond Zero Emissions, and others, and this cost can be verified against current solar thermal and wind power costs overseas.

The Australian Treasury Department modelling suggests that up to 50% of the emissions trading carbon abatement funds from 2015 to 2050 will be sent overseas. As the Australian Coalition party has pointed out, this grows to $58 Billion per annum in 2050. Over the 35 year period it totals around $780 Billion at Treasury's lower-end predicted international carbon price, growing to $131/ton in 2050. $780 Billion would otherwise fund 200% renewable energy for Australia if invested onshore.

At Treasury's higher-end prediction of the international carbon price in 2050 of $275/ton, the amount to be invested (sent) overseas totals around $1.5 Trillion, which would otherwise fund 400% renewable energy for Australia.

The analysis is easy to construct - a summary is provided below. It is based on the Treasury's prediction of overseas abatement growing to around 450MT per annum in 2050, due to inadequate onshore investment.

Picking Treasury's mid-point prediction of international carbon price, we should petition for 300% renewable energy in 35 years. This is then much like the Beyond Zero Emissions target of 100% in 10 years, but we keep going, and achieve 300% in 35 years.

One beauty of this model is that the Government can continue with its emissions trading scheme, almost exactly as planned, and with all of its inefficiencies that causes Treasury to currently predict Australia will only achieve 40% renewable energy in 2050.

The only thing that the Petition will request to change is that the carbon abatement that would be sent overseas is instead invested in Australia, with a focus on a coordinated plan for renewable energy, and a far-sighted plan for green energy export. With the second highest solar radiation of any country in the world (after Mongolia) Australia is very well placed to achieve this. The amount invested onshore need not be at any higher cost than the international carbon price, meaning there will be no additional cost to industry or consumers than currently planned.

Far-sighted planners in Europe are considering servicing Europe's power needs with solar energy from the Sahara. The reason for Australia to generate 200%, 300% or even 400% renewable energy is so that Australia can meet its own energy needs, and also export green energy to Asia and beyond. (See a later cost benefit analysis of this based on predictions of the cost of electricity in 2050).

Another cost benefit analysis of conversion to renewable energy is at my web site www.climate-change-explored-explained.com. It shows why an aggressive plan for 100% renewable energy is much cheaper than staying with coal. This is something that hardly anyone knows, because everyone just looks at the cost of renewable energy now. This analysis looks at the issue from a 30-40 year perspective, and it suggests we should be starting aggressively now.

Please do contact me if you are interested in creating and managing this formal petition, as I am also contacting other environmental groups, and we would best do this in a coordinated manner.

If you are unsure of what I am saying, I would instead be pleased if you would circulate what I suggesting for further discussion and comment.

Expected Cost of Overseas Permits

Australian Treasury Prediction of Low end and High End International Carbon Price

Cost of Carbon Abatement to be Sent Overseas						
	Low End Price			High End Price		
Year	Overseas Abatement MT	Carbon Price $/T	Investment $M	Overseas Abatement MT	Carbon Price $/T	Investment $M
2015	0	25	0	0	25	0
2016	13	28	360	13	32	413
2017	26	31	799	26	39	1010
2018	39	34	1315	39	46	1791
2019	51	37	1909	51	54	2755
2020	64	40	2581	64	61	3903
2021	77	43	3330	77	68	5235
2022	90	46	4158	90	75	6750
2023	103	49	5064	103	82	8449
2024	116	52	6047	116	89	10332
2025	129	55	7108	129	96	12398
2026	141	58	8247	141	104	14648
2027	154	61	9464	154	111	17082
2028	167	64	10759	167	118	19699
2029	180	67	12132	180	125	22500
2030	193	70	13583	193	132	25485
2031	206	73	15111	206	139	28653
2032	219	76	16718	219	146	32005
2033	231	80	18402	231	154	35541
2034	244	83	20164	244	161	39260
2035	257	86	22004	257	168	43163
2036	270	89	23922	270	175	47250
2037	283	92	25918	283	182	51520
2038	296	95	27991	296	189	55974
2039	309	98	30143	309	196	60612
2040	321	101	32372	321	204	65434
2041	334	104	34680	334	211	70439
2042	347	107	37065	347	218	75628
2043	360	110	39528	360	225	81000
2044	373	113	42069	373	232	86556
2045	386	116	44688	386	239	92296
2046	399	119	47384	399	246	98219
2047	411	122	50159	411	254	104327
2048	424	125	53011	424	261	110617
2049	437	128	55942	437	268	117092
2050	450	131	58950	450	275	123750
		Total ($M)	**$783,077**		**Total ($M)**	**$1,571,786**

Note to Reader

Frankly, I would not have been game to start pushing the idea of 200% or 400% renewable energy for Australia, here at this web site, if it were not for the Australian Government's recent carbon pricing details announcement, in which we intend to send $780B to overseas carbon offset schemes. Click on tab Policy Summary for more details of this. This is enough to fund 200% renewable energy for Australia. And indeed, at the Australian Treasury's higher prediction of the international 2050 carbon price of $275/ton, the amount sent overseas over 35 years doubles to $1.57 Trillion, which is more than enough for Australia to achieve 400% renewable energy !

I have pointed out that Australia could achieve 200%-400% renewable energy with 50% of the emissions trading funds (the amount expected to be sent overseas) to Rising Tide, the Australian Youth Climate Coalition, Senator Christine Milne, Beyond Zero Emissions and via this web site to you. The Australian Coalition won't point this out because they have no large-scale vision. I think the Greens are lost in the political maze.

It is probably hard for new readers to get their heads around 100%, 200% or 400% renewable energy. Such readers should look at the plans of Beyond Zero Emissions in order to see how easy it really is, and how it can meet our full year baseload and peak energy requirements. For the solar energy component of 100% renewable energy, which is the largest component, we just need 12 sites around regional Australia. For 200%, we would need 24 sites.

For Australia, the project would be like another Snowy Mountains scheme, that ran in the 1950's, created a lot of jobs, and was then probably the largest hydro prohect in the world. This new project for 100%, 200% or 400% renewable energy will take about 30 years to achieve. The book shows, that because of rising fossil fuel prices, it is in fact cheaper and more profitable for us if we start now.

A call for 300% renewable energy should be newsworthy, in Australia and the rest of the world. It is conservative in that it does not propose spending any more than proposed under the emissions trading proposal. Apart from that, it is obviously radical. It is actually a proposal for coordinated large-scale direct-action, using 50% of the carbon abatement funds.

The World Energy System for 2080

It is difficult to imagine how the world can cleanly generate its electrical energy needs without littering the landscape with wind towers.

The solution can be built as described here. Efforts from countries such as Australia can start to move towards this solution to be built over 50 years from 2030 to 2080, or sooner.

The world's total electricity requirements can be generated at 12 sites, as follows.

1. Four solar thermal sites, being Australia, the Sahara, North America and Mongolia.

2. Four offshore wind sites, being the southern ocean below Australia, below South America, and two sites off Africa.

3. Four offshore wave energy sites, again in the southern ocean below Australia, below South America, and two sites off Africa.

Each of these geographically dispersed sites will generate electrical energy for their local region and send excess energy to an international grid.

The system can be designed to minimise impact on land, and to create employment and income in continents such as Africa.

In this model, Australia could thus lead the generation of 25% of the world's energy needs. The discussion below describes further how Australia can be a very substantial energy exporter.

As a more conservative design, it also describes the expansion of Australia's 2050 300% energy system to generate 600% of its own energy needs by 2080, thus servicing 10% of the world's total energy needs, and at a cost similar to Australia's current emissions trading policy proposal.

Costing for the World Energy System for 2080

The book costed a system to meet the world's electricity requirements at $16.8 Trillion. Based on the population of 7 billion, this is $80 per head for 30 years. Current world electricity usage is 17,000 TWh p.a. This costing was based on near future predictions of photovoltiacs costs.

Below are two newer costings, both coming to $21 Trillion, involving solar thermal, wind and wave costings. In both cases the amount of equipment required is large and scarey. The first costing shows 8,000 large 220MW solar thermal facilities, 160,000 wind towers and 160,000 wave generators.

These numbers become less scarey if we think in terms of 50 years to build it all. It requires 160 large solar units to be built per year - this is though much more than we build now. It is also less scarey if we think that the numbers could halve, if in that time the ability to meet say 50% of energy requirements might be met locally by rooftop solar systems, new paint and window systems, or other local systems. It becomes more scarey again if we consider that the world's energy requirements will also double in that timeframe.

The first costing is based on the model described above, with four global sites each of solar thermal, wind and wave. Where this looks difficult for a country such as Australia to be one of the four global sites is the idea of having 2,000 large solar thermal facilities in our landscape.

World energy system 1								
Solar thermal								
		MWe/unit	GWh pa/unit	Units	GWe Total	TWh pa	Cost $M/MWe	Cost $B
Site 1		220	1400	2000	440	2800	10	4400
Site 2		220	1400	2000	440	2800	10	4400
Site 3		220	1400	2000	440	2800	10	4400
Site 4		220	1400	2000	440	2800	10	4400
Wind								
		MWe/unit	GWh pa/unit	Units	GWe Total	TWh pa	Cost $M/MWe	Cost $B
Site 1		7	30.66	40000	280	1226.4	2	560
Site 2		7	30.66	40000	280	1226.4	2	560
Site 2		7	30.66	40000	280	1226.4	2	560
Site 4		7	30.66	40000	280	1226.4	2	560
Wave								
		MWe/unit	GWh pa/unit	Units	GWe Total	TWh pa	Cost $M/MWe	Cost $B
Site 1		4	17.52	40000	160	700.8	2	320
Site 2		4	17.52	40000	160	700.8	2	320
Site 2		4	17.52	40000	160	700.8	2	320
Site 4		4	17.52	40000	160	700.8	2	320
Total						18908.8		21120

This first costing suggests that the contribution of wind and wave is quite small compared to the contribution of solar, based on the numbers used.

The second costing therefore explores an alternative model, whereby only solar thermal is used, across 10 global sites rather than just 4. That reduces the number of large facilities per site to 1200 each. In Australia, that could be implemented as 40 facilities at 30 sites. That is, 30 regional centres that are flanked by 10 facilities each on all four quarters. It is still a major commitment, but limits the effect on our landscape, and would allow Australia to produce 10% of the world's electricity requirements. This is an easier target for Australia to reach than the 25% target in the first model.

World energy system 2							
Solar thermal							
	MWe/unit	GWh pa/unit	Units	GWe Total	TWh pa	Cost $M/MWe	Cost $B
Site 1	220	1400	1200	264	1680	8	2112
Site 2	220	1400	1200	264	1680	8	2112
Site 3	220	1400	1200	264	1680	8	2112
Site 4	220	1400	1200	264	1680	8	2112
Site 5	220	1400	1200	264	1680	8	2112
Site 6	220	1400	1200	264	1680	8	2112
Site 7	220	1400	1200	264	1680	8	2112
Site 8	220	1400	1200	264	1680	8	2112
Site 9	220	1400	1200	264	1680	8	2112
Site 10	220	1400	1200	264	1680	8	2112
Total					16800		21120

Note: 1200 units at a site (country or region) can mean 30 spots where a cluster of 40 units are built.

For Australia, this compares to the BZE 10 year plan which proposes 12 spots (towns) with 10 units each.

1200 units is 10 times bigger, but is to be achieved in 70 years rather than just 10.

The cost to the world of $21 Trillion dollars, over say a 50 year build time, is still less than 1% of the world's annual GDP. One estimate of the full cost of the Iraq war is $4 Trillion, so it equates to one of those per decade.

The costing also suggests, that if the Australian Treasury modelling of our emissions trading proposal is correct, the $2 Trillion dollars to be spent onshore or sent overseas by 2050 could instead be used to meet 10% of the world's ongoing stationary energy needs, with Australia being one of the 10 sites proposed in the second costing model.

From Australia's perspective, the size of the effort to build 1200 solar thermal power stations is large. It is 24 facilities to build per year for 50 years or 17 per year for 70 years.

From a world perspective, the size of this effort is also large. However if this is what we need to do, then this is what we must do, starting now.

Profit to Australia from Generating 10% of World Electricity Requirements

Below is a high-level cost benefit analysis of Australia building 10% of the world's energy supply needs with renewable energy in 50 years. It shows $40B of expenditure each year for 50 years, totaling $2 Trillion. It shows income starting at a wholesale rate of $50/MWh (5 cents/Kwh) and growing 10% per year.

It shows that we are $230B in the red after 10 years, and but break even after 16 years.

It shows a $77 Trillion profit after 50 years (from our $2 Trillion investment) and $9 Trillion in income each year after that.

Bear in mind Australia's current GDP is about $1 Trillion p.a. at present, and what this vision might therefore mean for Australia.

An alternative analysis, with electricity prices rising more conservatively at only 6% per year to $2.70 per kw/h, but building the facilities over 70 years to 2080, achieves a similar end profit at $62 Trillion.

Further analyses, at price growth of 3% could be undertaken, although energy costs are likely to rise faster than that given the investments in renewable and fossil fuel energy that are being demanded by an increasingly affluent world.

In reality no-one can really predict what the price of electricity would be in 50 years. If the profit is not $62 Trillion or $77 Trillion, then what is it?

Year	Twh pa	Cost $b	$/mwh	Income $B	Position $B
1	34	40	50	2	-38
2	68	40	55	4	-75
3	102	40	61	6	-108
4	136	40	67	9	-139
5	170	40	73	12	-167
6	204	40	81	16	-190
7	238	40	89	21	-209
8	272	40	97	27	-223
9	306	40	107	33	-230
10	340	40	118	40	-230
11	374	40	130	49	-221
12	408	40	143	58	-203
13	442	40	157	69	-174
14	476	40	173	82	-132
15	510	40	190	97	-75
16	544	40	209	114	-1
17	578	40	230	133	91
18	612	40	253	155	206
19	646	40	278	180	346
20	680	40	306	208	514
21	714	40	336	240	714
22	748	40	370	277	951
23	782	40	407	318	1229
24	816	40	448	365	1554
25	850	40	492	419	1933
26	884	40	542	479	2372
27	918	40	596	547	2879
28	952	40	655	624	3463
29	986	40	721	711	4134
30	1020	40	793	809	4903
31	1054	40	872	920	5782
32	1088	40	960	1044	6787
33	1122	40	1056	1184	7931
34	1156	40	1161	1342	9233
35	1190	40	1277	1520	10714
36	1224	40	1405	1720	12393
37	1258	40	1546	1944	14298
38	1292	40	1700	2197	16454
39	1326	40	1870	2480	18894
40	1360	40	2057	2798	21652
41	1394	40	2263	3155	24767
42	1428	40	2489	3555	28281
43	1462	40	2738	4003	32245
44	1496	40	3012	4506	36711
45	1530	40	3313	5069	41740
46	1564	40	3645	5700	47400
47	1598	40	4009	6406	53766
48	1632	40	4410	7197	60923
49	1666	40	4851	8082	68965
50	1700	40	5336	9071	77996
		2000			

Each of the these spreadsheets can be obtained by contacting the author.

Explanation of the World Energy System Calculations

A. Mwe/unit is the size of a unit. Solar Reserve now plan 220 Mwe solar themal facilities. There is a promotional video on the web.

B. GWh pa/unit. The number of gigawatt hours per year such a unit can produce, based on specifications.

C. Units is a number of units required in order to achieve the planned over all sizing.

D. Gwe total is A X C/1000.

E. TWh pa is B X C /1000.

F. Cost is cost of such facilities per MWe. See discussion in the second letter of 19 April 2011.

G. Cost $B is total cost is A X C X F /1000.

Explanation of the 50 Year Profit Calculations

This is based on the previous World Energy Calculation where the capital cost to produce 17,000 Twh pa is $20T. For Australia to produce 10% of this 1700 Twh pa will cost $2T.

A. Twh pa grows to 1700 over 50 years.

B. Cost $B is $40B pa adding up to $2T over 50 years.

C. $/Mwh represents a wholesale cost of electricity, starting at $50/Mwh and growing 10% per annum.

D. Income $B is A X C /1000.

E. Position $B is the cumulative position of income minus cost.

Both of these calculations are based on current world electricity usage and current costs.

A costing technique used in this book is to assume that price reductions due to economies of scale or advances in technology will roughly offset increased growth in requirements/population.

The Journey

Below is a summary of how the ideas in the book and at this web site arose.

1. At the end of January 2010, I read the skeptic Ian Plimer's book that challenged climate science. I was fascinated by its arguments and it caused me to want to research the most likely truth.

2. After much further reading, I documented my research and my conclusions that climate science is well thought through and why the skeptical arguments appear wrong. That became Part A of the book. There was little similar analysis at the time.

3. I then moved on to analyse the proposed climate policies. Part B of the book provides an analysis of the Australian Government's proposed emissions trading scheme, the Australian Opposition party's direct-action proposal and overseas policies such as proposed in the USA. I came to the conclusion that both Australian proposals were inadequate. This caused me to propose an alternative large-scale direct-action plan for countries such as Australia and the USA to implement 100% renewable energy and also become carbon neutral in 30 years. The cost is $390B for Australia and $5.3T for the USA, over 30 years. Funding methods as simple at a charge of 5 cents per KW/h could cover the cost of the plan.

4. I also read climate change risk analyses from the IPCC and analyses for Australia. The risks go far beyond what most people realise. I became very stressed about discussing the detail in my book. In time, I found a way to discuss the extent of the risks, without discussing much of the detail.

5. The book, with its front cover subtitles, was intended to influence politicians, and copies were rushed out in July 2010 ahead of the Australian 2010 election. Further copies were sent post election, with updates discussing the effectiveness or not of a carbon tax which was a new policy under discussion.

6. In January 2011 I became aware of the work of Beyond Zero Emissions with their 10 year plan for Australia to implement 100% renewable energy. Their work is so good I considered whether there was value in me progressing further. I also analysed criticism of their work and did further background research into their cost calculations.

7. In February and March 2011 I explored the application of a similar costing model to my more conservative 30 year plan. I was astonished to see the prediction of $2 Trillion of savings from reduced coal usage over that time. The book update letter at "The Massive Savings from Reduced coal" resulted. It means that the alternative plan in the book does not have a cost - it is profitable.

8. In May 2011 I observed that there appeared to be little appetite for Australians to spend the small upfront $20B investment that would be required to achieve the $2 Trillion of savings (not that these savings are well understood) and so the next book update letter "It Need Not Cost us Anything" resulted. This letter explained that costs of feed-in tariffs could be borrowed, at a mere $20B over 10 years for Australia, so that both current and future generations could gain the environmental and financial benefits from a good renewable energy plan at no current increased cost of living.

9. In July 2011, after the details of the carbon pricing proposal were announced, I read the Australian Treasury papers and realised that, on the face of it, the amount of funding proposed to be sent overseas totalled between $780B and $1.5T over 35 years. The "Petition for 300% Renewable Energy" resulted. Although I have received some nice replies to my previous letters from politicians, I now instead started contacting respected national environmental groups, but have so far received no replies from them.

10. In August 2011, I am completing this project by exploring a world renewable energy system design. The "World Energy System for 2080" tab has resulted. It is the best world response, because each country's own response will be inadequate. The analysis shows the possible massive profits to Australia of providing 10% of the world energy needs, of possibly $77 Trillion over 50 years. To do so

requires multiplying the Beyond Zero Emissions plan for solar thermal energy by about 10, over 50 years, with 30 regional sites rather than just 12, each 4 times as big.

I conduct this project out of concern for our environment and our future. These days, I mainly argue the points on economic grounds. That seems the best way to possibly achieve a result.

If you like the information at this web site, then email its address to your local politician, your local environmental group and to some friends. If you like the information, send me an email - the address is on the home tab - I won't publish your address. If you disagree, let me know and I'll consider your comments. If the ideas get enough support, we can think about how to take them further.

I was encouraged by a reply from Barack Obama. I thus sent a further letter to him, via Australia's American Ambassador, and have also now sent these ideas for Australia to our new foreign affairs Minister, Bob Carr.

Letter to the American Ambassador

15 September 2011

Ambassador Jeffrey L. Bleich

The American Ambassador to the Commonwealth of Australia

The American Embassy

Moonah Place

Yarralumla ACT 2600

Dear Mr Bleich

Please find enclosed a gift from myself to your President Mr Barack Obama. It is the most recent edition of a book that I have written about the very important topic of climate change. I hope it can be provided to him as part of his current Australian visit.

It replaces and updates copies of an earlier edition I provided to you close to one year ago.

It describes why a rapid implementation of 100% renewable energy over 30 years will be cheaper for countries such as the USA and Australia than staying with coal.

It describes how this can be best achieved and funded, at no cost to current and future generations, and being cheaper than "business as usual".

The plan comes at lower cost than our Australian Government's emissions trading proposal and achieves a far greater outcome. The plans make Australia or the USA carbon neutral or even negative in 30 years. This compares with the Australian Government's plan which achieves next to no net onshore emissions reduction from current levels in the same timeframe, but comes at far greater cost.

This is an important book to provide to the President. I trust he will find it an interesting read.

Yours sincerely

Stephen Harvie

Replies to Letters

Following are some of the replies received from copies of the book and associated letters sent to various politicians and other people.

Replies of various natures were received from Barack Obama, Christine Milne, Kevin Rudd, Greg Combet, Greg Hunt, Rob Oakeshott, Wayne Swan's office, Tony Abbott's office, Peter Garrett's office, Climate Commissioner Tim Flanney's office, and a letter from Ambassador Jeffery Bleich to accompany the letter from Barack Obama.

Rob Oakeshott and Greg Combet went to some effort in their replies to explain proposed Government policy.

SENATOR CHRISTINE MILNE
Australian Greens Senator for Tasmania

Mr. Stephen Harvie

Dear Stephen,

Thank you very much for sending Bob Brown a copy of your book *Climate Change Explored and Explained.*

Your letter to Senator Bob Brown was forwarded to my office as I am the Australian Greens spokesperson for Climate Change, Energy and Transport.

Australia needs people like you who, in an unbiased manner, explore the issues and then come to the conclusion that man-made climate change is one of the biggest problems mankind is facing. Your policy suggestions and proposals for action are informative for anyone who is seriously concerned about the future of mankind and our planet.

Late last year the Australian Greens launched their *Safe Climate Bill*, the first comprehensive legislative attempt to set out what we need to achieve to deliver a safe climate to our children.

The *Safe Climate Bill* outlines how an effective emissions trading scheme for Australia would work, while creating jobs, not losing them. I enclose our media release on the subject for your reference.

I hope your book will reach a wide audience.

Yours Sincerely

Christine Milne
Australian Greens Senator for Tasmania
www.christinemilne.org.au

Parliament House, Canberra ACT 2600 Australia • Phone: 02 6277 3063 • Fax: 02 6277 5720
Hobart Electorate Office, GPO Box 896, Hobart Tasmania 7001 Australia • Phone: 03 6224 8899 • Fax: 03 6224 7599
Email: senator.milne@aph.gov.au
www.christinemilne.org.au

Australian Government

Department of Foreign Affairs and Trade

2 December 2010

Mr Stephen Harvie

Dear Mr Harvie

The Minister for Foreign Affairs, the Hon Kevin Rudd MP, has asked me to thank you for your gift of your book 'Climate Change Explored and Explained' enclosed with your letter dated 3 November 2010.

You are to be congratulated for your efforts in producing such a comprehensive, well-researched and well-argued contribution to the debate on climate change. Your book will be valuable in informing the development of Australia's foreign policy on this issue.

Yours sincerely

Climate Change and Environment Section

R G Casey Building John McEwen Cres Barton 0221
Telephone:02 6261 1111 Facsimile:02 6261 3111

THE WHITE HOUSE

WASHINGTON

February 6, 2012

Mr. Stephen Harvie

AUSTRALIA

Dear Stephen:

I would like to extend my deepest thanks for your kind gift. Your thoughtfulness and generosity are much appreciated.

As our world grows increasingly interdependent, I look forward to working together to the benefit of our two nations, and to strengthening the bonds between our people.

Thank you, again, for the wonderful gift. I wish you all the best.

Sincerely,

A Final Closing Letter

A range of commentators are now explaining that renewable energy is now cheaper than staying with coal and/or that a carbon price will not be sufficient to bring about the required level of climate action in the necessary timeframes.

The range of proposals in this book, of 100% renewable energy, 300% renewable energy or 600% renewable energy for Australia (10% of world) and 100% renewable energy for the world represent visions that we need to start working aggressively toward.

The following letter has been sent to Ambassador Jeffrey Bleich of Australia's American embassy, and similar letters have been sent to Senator Christine Milne, The Hon Kevin Rudd and the Hon Malcolm Turnbull, as some of the best thinkers across the key Australian political parties.

Further such letters, and further electronic copies of this book will continue to be sent by this author, or if you like, by yourself.

If you would like to make suggestions or other commentary about the ideas in this book or how to promote them, do contact the author directly or via email address c-c-e-e@grapevine.com.au

Sixth Letter

2 April 2012

Ambassador Jeffrey Bleich

The American Ambassador to the Commonwealth of Australia

The American Embassy

Moonah Place

Yarralumla ACT 2600

Dear Mr Bleich

Thank you for your letter of 5 March 2012 forwarding the note from the White House thanking me for the gift of my book Climate Change Explored and Explained.

I would like to provide you with this latest edition of this book on the enclosed CD. It is an ebook. It presents well in iBooks on an iPad. It is also viewable on a desktop computer.

The main update to the book is the inclusion at the back of the book's web site material.

This includes a high level design and costing for a **world 100% renewable energy system**, in which Australia plays a major part in generating 10% of the world's renewable energy needs.

I do believe this proposed world renewable energy system needs to be built, as each country cannot solve this problem alone.

Participation in this world system should be open to any country that is either capable of generating more than its own energy requirements, or is willing to participate in joint venture arrangements with other countries that can, like Australia. Australia and the USA could work together to begin this process.

A similar proposal has been produced by the World Wildlife Foundation and Ecofys. ***The Energy Report - 100% Renewable Energy by 2050***, describes how an international 100% renewable energy grid would be constructed. I do think my proposed design approach has advantages and is more attainable.

The WWF also describes a plan for Australia for 2050 as part an overall world plan and how ***Australia could lead with the world's cheapest renewable energy***.

I would encourage you would look at these ideas and consider what might be done to help bring them about.

Yours sincerely

Stephen Harvie

www.ingramcontent.com/pod-product-compliance
Lightning Source LLC
LaVergne TN
LVHW070953210926
848482LV00002B/205
* 9 7 8 0 6 4 6 5 3 5 1 3 5 *